BUSINESS/SCIENCE/TECHNOLOGY DIVISION
CHICAGO PUBLIC LIBRARY
400 SOUTH STATE STREET
CHICAGO, IL 60605

D0660737

Atomic Physics And Human Knowledge

Niels Bohr

DOVER PUBLICATIONS, INC.
Mineola, New York

Bibliographical Note

This Dover edition, first published in 2010, is an unabridged republication of the work originally published in 1961 by Science Editions, Inc., New York.

Library of Congress Cataloging-in-Publication Data

Bohr, Niels, 1885–1962.
  [Atomfysik og menneskelig erkenselse. English.]
  Atomic physics and human knowledge / niels Bohr.
     p. cm.
  "This Dover edition, first published in 2010, is an unabridged republication of the work originally published in 1961 by Science Editions, In., New York"—Prelim.
  ISBN-13: 978-0-486-47928-6
  ISBN-10: 0-486-47928-5
     1. Physics—Philosophy. 2. Nuclear physics. 3. Knowledge, Theory of. I. Title.

QC6.B598 2010
530.01—dc22

2010022935

Manufactured in the United States by Courier Corporation
47928501
www.doverpublications.com

R0432887513

CHICAGO PUBLIC LIBRARY

Preface

This collection of articles, written on various occasions within the last 25 years, forms a sequel to earlier essays edited by the Cambridge University Press, 1934, in a volume titled *Atomic Theory and the Description of Nature*. The theme of the papers is the epistemological lesson which the modern development of atomic physics has given us and its relevance for analysis and synthesis in many fields of human knowledge. The articles in the previous edition were written at a time when the establishment of the mathematical methods of quantum mechanics had created a firm foundation for the consistent treatment of atomic phenomena, and the conditions for an unambiguous account of experience within this framework were characterized by the notion of complementarity. In the papers collected here, this approach is further developed in logical formulation and given broader application. Of course, much repetition has been unavoidable, but it is hoped that this may serve to illustrate the gradual clarification of the argumentation, especially as regards more concise terminology.

In the development of the views concerned, discussions with former and present collaborators at the Institute for Theoretical Physics in the University of Copenhagen have been most valuable to me. For assistance in the elaboration of the articles in this volume, I am especially indebted to Oskar Klein and Léon Rosenfeld, now in the universities of Stockholm and Manchester, as well as to Stefan Rozental and Aage Petersen at the Copenhagen Institute. Also I should like to extend my thanks to Mrs. S. Hellmann for her most effective help in the preparation of the articles and the present edition.

NIELS BOHR

Copenhagen
August 1957

Contents

Atomic Physics And Human Knowledge

Introduction

The importance of physical science for the development of general philosophical thinking rests not only on its contributions to our steadily increasing knowledge of that nature of which we ourselves are part, but also on the opportunities which time and again it has offered for examination and refinement of our conceptual tools. In our century, the study of the atomic constitution of matter has revealed an unsuspected limitation of the scope of classical physical ideas and has thrown new light on the demands on scientific explanation incorporated in traditional philosophy. The revision of the foundation for the unambiguous application of our elementary concepts, necessary for comprehension of atomic phenomena, therefore has a bearing far beyond the special domain of physical science.

The main point of the lesson given us by the development of atomic physics is, as is well known, the recognition of a feature of wholeness in atomic processes, disclosed by the discovery of the quantum of action. The following articles present the essential aspects of the situation in quantum physics and, at the same time, stress the points of similarity it exhibits to our position in other fields of knowledge beyond the scope of the mechanical conception of nature. We are not dealing here with more or less vague analogies, but with an investigation of the conditions for the proper use of our conceptual means of expression. Such considerations not only aim at making us familiar with the novel situation in physical science, but might on account of the comparatively simple character of atomic problems be helpful in clarifying the conditions for objective description in wider fields.

Although the seven essays here collected are thus closely interconnected, they fall into three separate groups originating from the years 1932–1938, 1949, and 1954–1957, respectively. The first three papers, directly related to the articles in the previous edition, discuss biological and anthropological problems referring to the features of wholeness presented by living organisms and human cultures. Of course, it is in no way attempted to give an exhaustive treatment of these topics, but only to indicate how the problems present themselves against the background of the general lesson of atomic physics.

The fourth article deals with the discussion among physicists of the epistemological problems raised by quantum physics. Owing to the character of the topic, some reference to the mathematical tools has been unavoidable, but the understanding of the arguments demands no special knowledge. The debate led to a clarification of the new aspects of the observational problem, implied by the circumstance that the interaction between atomic objects and measuring instruments forms an integral part of quantum phenomena. Therefore, evidence gained by different experimental arrangements cannot be comprehended on accustomed lines, and the necessity of taking into account the conditions under which experience is obtained calls directly for the complementary mode of description.

The last group of articles is closely related to the first, but it is hoped that the improved terminology used to present the situation in quantum physics has made the general argument more easily accessible. In its application to problems of broader scope, emphasis is laid especially on the presuppositions for unambiguous use of the concepts employed in the account of experience. The gist of the argument is that for objective description and harmonious comprehension it is necessary in almost every field of knowledge to pay attention to the circumstances under which evidence is obtained.

Light and Life

1932

As a physicist whose studies are limited to the properties of inanimate bodies, it is not without hesitation that I have accepted the kind invitation to address this assembly of scientists met together to forward our knowledge of the beneficial effects of light in the cure of diseases. Unable as I am to contribute to this beautiful branch of science that is so important for the welfare of mankind, I could at most comment on the purely inorganic light phenomena which have exerted a special attraction on physicists throughout the ages not least owing to the fact that light is our principal tool of observation. I have thought, however, that on this occasion it might perhaps be of interest in such a comment to enter on the problem of how far the results reached in the more limited domain of physics may influence our views as regards the position of living organisms within the general edifice of natural science. Notwithstanding the subtle character of the riddles of life, this problem has presented itself at every stage of science, the very essence of scientific explanation being the analysis of more complex phenomena into simpler ones. At the moment it is the essential limitation of the mechanical description of natural phenomena revealed by the recent development of atomic theory which has lent new interest to the old problem. This development originated just in the closer study of the interaction between light and material bodies which presents features that defeat certain demands hitherto considered as indispensable in a physical explanation. As I shall endeavour to show, the efforts of physicists to master this situation resemble in some way the attitude towards the aspects of life always taken more or less intuitively by biologists. Still, I wish to stress at once that it is only in this formal respect that light, which is perhaps the least complex of all physical phenomena, exhibits an analogy to life which shows a diversity beyond the grasp of scientific analysis.

From a physical standpoint, light may be defined as transmission of energy between material bodies at a distance. As is well known, such effects find a simple explanation within the electromagnetic theory which may be regarded as a rational extension of classical mechanics suited to alleviate the contrast between action at a distance and at contact. According to this theory, light is described as coupled electric and magnetic oscillations differing from ordinary electromagnetic waves of radio transmission only by the greater frequency of vibration and the smaller wave-length. In fact, the practically rectilinear propagation of light, on which rests the location of bodies by direct vision or by suitable optical instruments, depends entirely on the smallness of the wave-length compared with the dimensions of the bodies concerned and of the instruments. At the same time, the wave character of light propagation not only forms the basis for our account of colour phenomena, which in spectroscopy have yielded such important information of the constitution of material bodies, but is also essential for every refined analysis of optical phenomena. As a typical example, I need only mention the interference patterns which appear when light from one source can travel to a screen along two different paths. Here we find that the effects which would be produced by the separate light beams are strengthened at such points of the screen where the phases of the two wave trains coincide, that is, where the electric and magnetic oscillations in the two beams have the same directions, while the effects are weakened and may even disappear at points where these oscillations have opposite directions and where the wave trains are said to be out of phase with one another. These interference patterns offer so thorough a test of the wave picture of light propagation that this picture cannot be considered as a hypothesis in the usual sense of this word, but may rather be regarded as the adequate account of the phenomena observed.

Still, as you all know, the problem of the nature of light has been subjected to renewed discussion in recent years, on account of the discovery of an essential feature of atomicity in the mechanism of energy transmission which is quite unintelligible from the point of view of the electromagnetic theory. In fact, any energy transfer by light can be traced down to individual processes in each of which a so-called light quantum is exchanged whose energy is equal to the product of the frequency of the electromagnetic oscillations and the universal quantum of action or Planck's constant. The obvious contrast between this atomicity of the light effect and the continuity of the energy transfer in the electromagnetic theory presents us with a dilemma of a character hitherto unknown in physics. Thus, in spite of its obvious insufficiency, there can be no question of replacing the wave picture of light propagation by some other picture leaning on ordinary mechanical ideas. Especially, it should be emphasized that light quanta cannot be regarded as particles to which a well-defined path in the sense of ordinary mechanics can be ascribed. Just as an interference pattern would completely disappear if, in order to make sure that the light energy travelled only along one of the two paths between the source and the screen, we would stop one of the beams by a non-transparent body, so is it impossible in any phenomenon for which the wave constitution of light is essential to trace the path of the individual light quanta without essentially disturbing the phenomenon under investigation. Indeed, the spatial continuity of our picture of light propagation and the atomicity of the light effects are complementary aspects in the sense that they account for equally important features of the light phenomena which can never be brought into direct contradiction with one another, since their closer analysis in mechanical terms demand mutually exclusive experimental arrangements. At the same time, this very situation forces us to renounce on a complete causal account of the light phenomena and to be content with probability laws based on the fact that the electromagnetic description of energy transfer remains valid in a statistical sense. This forms a typical application of the so-called correspondence argument which expresses the endeavour of utilizing to the outmost extent the concepts of the classical theories of mechanics and electrodynamics, in spite of the contrast between these theories and the quantum of action.

At first, this situation may appear very uncomfortable but, as has often happened in science when new discoveries have led to the recognition of an essential limitation of concepts hitherto considered as indispensable, we are rewarded by getting a wider view and a greater power to correlate phenomena which before might even have appeared as contradictory. Indeed, the limitation of classical mechanics symbolized by the quantum of action has offered a clue to our understanding of the intrinsic stability of atoms on which the mechanical description of natural phenomena is essentially based. Of course, it has always been a fundamental feature of the atomic theory that the indivisibility of the atoms cannot be understood in mechanical terms, and this situation remained practically unchanged even after the indivisibility of atoms was replaced by that of the elementary electric particles, electrons and protons, of which atoms and molecules are built up. What I am referring to is not the problem of the intrinsic stability of these elementary particles but that of the atomic structures composed of them. If we attack this problem from the point of view of mechanics or of the electromagnetic theory, we find no sufficient basis on which to account for the specific properties of the elements and not even for the existence of rigid bodies on which all measurements used for ordering phenomena in space and time ultimately rest. These difficulties are now overcome by the recognition that any well-defined change of an atom is an individual process consisting in a complete transition of the atom from one of its so-called stationary states to another. Moreover, since just one light quantum is exchanged in a transition process by which light is emitted or absorbed by an atom, we are able by means of spectroscopic observations to measure directly the energy of each of these stationary states. The information thus derived has also been most instructively corroborated by the study of the energy exchanges which take place in atomic collisions and in chemical reactions.

In recent years a remarkable development of atomic mechanics along the lines of the correspondence argument has taken place, affording us with proper methods of calculating the energies of the stationary states of atoms and the probabilities of transition processes, thus making our account of atomic properties as comprehensive as the coordination of astronomical experience by Newtonian mechanics. Notwithstanding the greater complexity of the general problems of atomic mechanics, the lesson taught us by the analysis of the simpler light effects has been most important for this development. Thus, the unambiguous use of the concept of stationary states stands in a similar relation of complementarity to a mechanical analysis of intra-atomic motions as do light quanta to the electromagnetic theory of radiation. Indeed, any attempt to trace the detailed course of a transition process would involve an uncontrollable exchange of energy between the atom and the measuring instruments which would completely disturb the very energy balance we set out to investigate. The causal mechanical coordination of experience can be accomplished only in cases where the action involved is large compared with the quantum and where, therefore, a subdivision of the phenomena is possible. If this condition is not fulfilled, the action of the measuring instruments on the object under investigation cannot be disregarded and will entail a mutual exclusion of the various kinds of information required for a complete mechanical description of the usual type. This apparent incompleteness of the mechanical analysis of atomic phenomena issues ultimately from the ignorance of the reaction of the object on the measuring instruments inherent in any measurement. Just as the general concept of relativity expresses the essential dependence of any phenomenon on the frame of reference used for its coordination in space and time, the notion of complementarity serves to symbolize the fundamental limitation, met with in atomic physics, of the objective existence of phenomena independent of the means of their observation.

This revision of the foundations of mechanics, extending to the very idea of physical explanation, not only is essential for the full appreciation of the situation in atomic theory but also creates a new background for the discussion of the problems of life in their relation to physics. In no way does this mean that in atomic phenomena we meet with features which show a closer resemblance to the properties of living organisms than do ordinary physical effects. At first sight, the essentially statistical character of atomic mechanics might even seem to conflict with the marvellously refined organisation of living beings. We must keep in mind, however, that just this complementary mode of description leaves room for regularities in atomic processes foreign to mechanics but as essential for our account of the behaviour of living organisms as for the explanation of the specific properties of inorganic matter. Thus, in the carbon assimilation of plants, on which so largely depends also the nourishment of animals, we are dealing with a phenomenon for the understanding of which the individuality of photo-chemical processes is clearly essential. Likewise, the non-mechanical stability of atomic structures is markedly exhibited in the characteristic properties of such highly complicated chemical combinations as chlorophyll or hemoglobine which play a fundamental part in the mechanism of plant assimilation and animal respiration. Still, analogies from ordinary chemical experience, like the ancient comparison of life with fire, will of course yield no more satisfactory explanation of living organisms than will their resemblance with such purely mechanical contrivances as a clock-work. Indeed, the essential characteristics of living beings must be sought in a peculiar organisation in which features that may be analyzed by usual mechanics are interwoven with typically atomistic features to an extent unparalleled in inanimate matter.

An instructive illustration of the degree to which this organisation is developed is exhibited by the construction and function of the eye, for the exploration of which the simplicity of light phenomena have again been most helpful. I need not here go into details but shall just remind you how ophthalmology has revealed to us the ideal properties of the human eye as an optical instrument. Indeed, the limit imposed on the image formation by the unavoidable interference effects coincides practically with the size of such partitions of the retina which have separate nervous connection with the brain. Moreover, since the absorption of a single light quantum by each of these retinal partitions is sufficient for a sight impression, the sensitiveness of the eye may be said to have reached the limit set by the atomic character of the light processes. The efficiency of the eye in both of these respects is actually the same as that obtained in a good telescope or microscope connected with a suitable amplifier so as to make the individual processes observable. It is true that it is possible by such instruments to essentially increase our powers of observation, but, due to the limits imposed by the fundamental properties of the light phenomena, no instrument is imaginable which is more efficient for its purpose than the eye. Now, this ideal refinement of the eye, recognized through the recent development of physics, suggests that also other organs, whether they serve for the reception of information from the surroundings or for the reaction to sense impressions, will exhibit a similar adaptation to their purpose, and that also here the feature of individuality symbolized by the quantum of action is of decisive importance in connection with some amplifying mechanism. That it has been possible to trace this limit in the eye but not, so far, in any other organ is due simply to the extreme simplicity of the light phenomena to which we have referred before.

The recognition of the essential importance of atomistic features in the mechanism of living organisms is in no way sufficient, however, for a comprehensive explanation of biological phenomena. The question at issue, therefore, is whether some fundamental traits are still missing in the analysis of natural phenomena before we can reach an understanding of life on the basis of physical experience. Notwithstanding the fact that the multifarious biological phenomena are practically inexhaustible, an answer to this question can hardly be given without an examination of the meaning to be given to physical explanation still more penetrating than that to which the discovery of the quantum of action has already forced us. On the one hand, the wonderful features which are constantly revealed in physiological investigations and which differ so markedly from what is known of inorganic matter have lead biologists to the belief that no proper understanding of the essential aspects of life is possible in purely physical terms. On the other hand, the view known as vitalism can hardly be given an unambiguous expression by the assumption that a peculiar vital force, unknown to physics, governs all organic life. Indeed, I think we all agree with Newton that the ultimate basis of science is the expectation that nature will exhibit the same effects under the same conditions. If, therefore, we were able to push the analysis of the mechanism of living organisms as far as that of atomic phenomena, we should not expect to find any features foreign to inorganic matter. In this dilemma it must be kept in mind, however, that the conditions in biological and physical research are not directly comparable, since the necessity of keeping the object of investigation alive imposes a restriction on the former which finds no counterpart in the latter. Thus, we should doubtless kill an animal if we tried to carry the investigation of its organs so far that we could tell the part played by the single atoms in vital functions. In every experiment on living organisms there must remain some uncertainty as regards the physical conditions to which they are subjected, and the idea suggests itself that the minimal freedom we must allow the organism will be just large enough to permit it, so to say, to hide its ultimate secrets from us. On this view, the very existence of life must in biology be considered as an elementary fact, just as in atomic physics the existence of the quantum of action has to be taken as a basic fact that cannot be derived from ordinary mechanical physics. Indeed, the essential non-analyzability of atomic stability in mechanical terms presents a close analogy to the impossibility of a physical or chemical explanation of the peculiar functions characteristic of life.

In tracing this analogy, however, we must remember that the problems present essentially different aspects in atomic physics and in biology. While in the former field we are primarily interested in the behaviour of matter in its simplest forms, the complexity of the material systems with which we are concerned in biology is of a fundamental nature, since even the most primitive organisms contain large numbers of atoms. It is true that the wide field of application of ordinary mechanics, including our account of the measuring instruments used in atomic physics, rests just on the possibility of largely disregarding the complementarity of the description entailed by the quantum of action in cases where we are dealing with bodies containing a great number of atoms. Notwithstanding the essential importance of the atomistic features, it is typical of biological research, however, that we can never control the external conditions to which any separate atom is subjected to the extent possible in the fundamental experiments of atomic physics. In fact, we cannot even tell which particular atoms really belong to a living organism, since any vital function is accompanied by an exchange of material through which atoms are constantly taken up into and expelled from the organisation which constitutes the living being. Indeed, this exchange of matter extends to all parts of a living organism to a degree which prevents a sharp distinction on an atomic scale between those features of its mechanism which can be unambiguously accounted for on usual mechanics and those for which a regard of the quantum of action is decisive. This fundamental difference between physical and biological research implies that no well-defined limit can be drawn for the applicability of physical ideas to the problems of life which corresponds to the distinction between the field of causal mechanical description and proper quantum phenomena in atomic mechanics. This apparent limitation of the analogy in question is rooted in the very definitions of the words life and mechanics which are ultimately a matter of convenience. On the one hand, the question of a limitation of physics in biology would lose any meaning if, instead of distinguishing between living organisms and inanimate bodies, we extended the idea of life to all natural phenomena. On the other hand, if, in accordance with common language, we were to reserve the word mechanics for the unambiguous causal description of natural phenomena, such a term as atomic mechanics would become meaningless. I shall not enter further into such purely terminological points but only add that the essence of the analogy being considered is the obvious exclusiveness between such typical aspects of life as the self-preservation and the self-generation of individuals, on the one hand, and the subdivision necessary for any physical analysis on the other hand. Owing to this essential feature of complementarity, the concept of purpose, which is foreign to mechanical analysis, finds a certain field of application in biology. Indeed, in this sense teleological argumentation may be regarded as a legitimate feature of physiological description which takes due regard to the characteristics of life in a way analogous to the recognition of the quantum of action in the correspondence argument of atomic physics.

In discussing the applicability of purely physical ideas to living organisms we have, of course, treated life just as any other phenomenon of the material world. I need hardly emphasize, however, that this attitude, which is characteristic of biological research, involves no disregard of the psychological aspect of life. On the contrary, the recognition of the limitation of mechanical concepts in atomic physics would rather seem suited to conciliate the apparently contrasting viewpoints of physiology and psychology. Indeed, the necessity of considering the interaction between the measuring instruments and the object under investigation in atomic mechanics exhibits a close analogy to the peculiar difficulties in psychological analysis arising from the fact that the mental content is invariably altered when the attention is concentrated on any special feature of it. It will carry us too far from our subject to enlarge upon this analogy which offers an essential clarification of the psycho-physical parallelism. However, I should like to emphasize that considerations of the kind here mentioned are entirely opposed to any attempt of seeking new possibilities for a spiritual influence on the behaviour of matter in the statistical description of atomic phenomena. For instance, it is impossible, from our standpoint, to attach an unambiguous meaning to the view sometimes expressed that the probability of the occurrence of certain atomic processes in the body might be under the direct influence of the will. In fact, according to the generalized interpretation of the psycho-physical parallelism, the freedom of the will is to be considered as a feature of conscious life which corresponds to functions of the organism that not only evade a causal mechanical description but resist even a physical analysis carried to the extent required for an unambiguous application of the statistical laws of atomic mechanics. Without entering into metaphysical speculations, I may perhaps add that an analysis of the very concept of explanation would, naturally, begin and end with a renunciation as to explaining our own conscious activity.

In conclusion, I need hardly emphasize that with none of my remarks have I intended to express any kind of scepticism as to the future development of physical and biological sciences. Such scepticism would, indeed, be far from the mind of physicists at the present time when just the recognition of the limited character of our most fundamental concepts has resulted in such a remarkable development of our science. Nor has the renunciation of an explanation of life impeded the wonderful progress which has taken place in all branches of biology, including those which have proved so beneficial in the art of medicine. Even if we cannot make a sharp physical distinction between health and disease, there is surely no room for scepticism in the special field which is the subject of this congress, as long as one does not leave the highroad of progress which has been followed with so great success since the pioneer work of Finsen, and the distinguishing mark of which is the most intimate combination of the investigation of the medical effects of light therapy with the study of its physical aspects.

Biology
and
Atomic Physics

1937

The immortal work of Galvani which inaugurated a new epoch in the whole field of science is a most brilliant illustration of the extreme fruitfulness of an intimate combination of the exploration of the laws of inanimate nature with the study of the properties of living organisms. At this occasion, it may therefore be appropriate to review the attitude which scientists through the ages have taken to the question of the relationship beween physics and biology and especially to discuss the outlook created in this respect by the extraordinary development of atomic theory in recent time.

From the very dawn of science, atomic theory has indeed been at the focus of interest in connection with the efforts to attain a comprehensive view of the great variety of natural phenomena. Thus already Democritus, who with so deep intuition emphasized the necessity of atomism for any rational account of the ordinary properties of matter, attempted, as is well known, also to utilize atomistic ideas for the explanation of the peculiarities of organic life and even of human psychology. In view of the fantastic character of such extreme materialistic conceptions, it was a natural reaction when Aristotle, with his masterly comprehension of the knowledge of his time in physics as well as in biology, rejected atomic theory entirely and tried to provide a sufficiently broad frame for an account of the wealth of natural phenomena on the basis of essentially teleological ideas. The exaggeration of the Aristotelian doctrine, on its side, was, however, clearly brought to light by the gradual recognition of elementary laws of nature valid as well for inanimate bodies as for living organisms.

When thinking of the establishment of the principles of mechanics, which were to become the very foundations of physical science, it is, in this connection, not without interest to realize that Archimedes' discovery of the principle of equilibrium of floating objects, which, according to a familiar tradition, was suggested to him by the sensation of uplift of his own body in a bath tub, might just as well have been based on common experience regarding the loss of weight of stones in water. Likewise it is to be regarded as quite accidental that Galileo was led to the recognition of the fundamental laws of dynamics by observing the pendulum motion of a chandelier in the beautiful cathedral of Pisa, and not by looking at a child in a swing. Yet such purely external analogies were, of course, only of little weight for the growing appreciation of the essential unity of the principles governing natural phenomena, as compared to the deep-rooted similarities between living organisms and technical machinery that were disclosed by the studies of anatomy and physiology, pursued so intensely at the time of the Renaissance especially here in Italy.

The enthusiasm for the prospects opened by the success of the new experimental approach to natural philosophy—encouraged in equal manner by the widening of the world picture due to the vision of Copernicus and by the elucidation of circulation mechanisms in animal bodies, initiated by Harvey's great achievement—found perhaps its most striking expression in the work of Borelli, who succeeded to clarify in so fine detail the mechanical function of skeleton and muscles in animal motion. The classical character of this work is in no way impeded by attempts of Borelli himself and his followers also to explain nervous action and glandular secretion by means of primitive mechanical models, the obvious arbitrariness and coarseness of which soon gave rise to general criticism, still remembered by the semi-ironical name of "iatro-physicists" attached to the Borellian school. Likewise the endeavours, sound in their root, to apply the growing knowledge of typically chemical transformations of matter to physiological processes, which found so enthusiastic an exponent in Sylvius, rapidly led, by exaggerations of superficial resemblances of digestion and fermentation with the simplest inorganic reactions and their rash application to medical purposes, to an opposition which has found its expression in the labelling of such premature endeavours as "iatro-chemistry."

To us the reasons for the shortcomings of these pioneer efforts to utilize physics and chemistry for a comprehensive explanation of the properties of living organisms are evident. Not only had one to wait until Lavoisier's time for the disclosure of the elementary principles of chemistry, which were to give the clue to the understanding of respiration and later to provide the basis for the extraordinary development of so-called organic chemistry, but, before Galvani's discoveries, a whole fundamental aspect of the laws of physics lay still hidden. It is most suggestive to think that the germ which, in the hands of Volta, Oersted, Faraday, and Maxwell, was to develop into a structure rivalling Newtonian mechanics in importance grew out of researches with a biological aim. In fact, it is difficult to imagine that the progress from experiments with electrically charged bodies, however fruitful in Franklin's hands, to the study of galvanic currents could have been achieved if the sensitive instruments necessary for the detection of such currents, afterwards so readily constructed, had not been provided by nature itself in the nervous fabric of higher animals.

It is impossible here to sketch, even in outline, the tremendous development of physics and chemistry since the days of Galvani, or to enumerate the discoveries in all branches of biology in the last century. We need only recall the lines leading from the pioneer work, in this venerable university, of Malpighi and Spallanzani to modern embryology and bacteriology respectively, or from Galvani himself to the recent fascinating researches on nerve impulses. In spite of the far-reaching understanding, thus obtained, of the physical and chemical aspect of many typical biological reactions, the marvellous fineness of structure of the organisms and their wealth of interconnected regulation mechanisms go still so far beyond any experience about inanimate nature that we feel as removed as ever from an explanation of life itself on such lines. Indeed, when we witness the passionate scientific controversies as regards the bearing on this problem of the recent discoveries of poisoning effects and generative properties of so-called virus, we find ourselves presented with a dilemma just as acute as that with which Democritus and Aristotle were confronted.

In this situation it is again upon atomic theory that interest is concentrated, although against a very different background. Not only has this theory, since Dalton applied with such decisive success atomistic conceptions to the elucidation of the quantitative laws governing the constitution of chemical compounds, become the indispensable foundation and never-failing guide of all reasoning in chemistry; but the wonderful refinement of experimental technique in physics has even given us the means of studying phenomena which directly depend on the action of individual atoms. At the same time that this development has thus removed the last traces of the traditional prejudice that, due to the coarseness of our senses, any proof of the actual existence of atoms would forever remain beyond the reach of human experience, it has revealed still deeper features of atomicity in the laws of nature than those expressed by the old doctrine of the limited divisibility of matter. We have indeed been taught that the very conceptual frame, appropriate both to give account of our experience in everyday life and to formulate the whole system of laws applying to the behaviour of matter in bulk and constituting the imposing edifice of so-called classical physics, had to be essentially widened if it was to comprehend proper atomic phenomena. In order to appreciate the possibilities which this new outlook in natural philosophy provides with respect to a rational attitude towards the fundamental problems of biology, it will, however, be necessary to recall briefly the principal lines of the development which has led to the elucidation of the situation in atomic theory.

The starting-point of modern atomic physics was, as is well known, the recognition of the atomic nature of electricity itself, first indicated by Faraday's famous researches on galvanic electrolysis and definitely established by the isolation of the electron in the beautiful phenomena of electric discharges through rarefied gases, which attracted so much attention towards the end of the last century. While J. J. Thomson's brilliant researches soon brought to light the essential part played by electrons in most varied physical and chemical phenomena, our knowledge of the structural units of matter was, however, not completed until Rutherford's discovery of the atomic nucleus, crowning his pioneer work on the spontaneous radioactive transmutations of certain heavy elements. Indeed, this discovery offered for the first time an unquestionable explanation of the invariability of the elements in ordinary chemical reactions, in which the minute heavy nucleus remains unaltered, while only the distribution of the light electrons around it is affected. Moreover, it provides an immediate understanding not only of the origin of natural radioactivity, in which we witness an explosion of the nucleus itself, but also of the possibility, subsequently discovered by Rutherford, of inducing transmutations of elements by bombardment with high-speed heavy particles which, in colliding with the nuclei, may cause their disintegration.

It would carry us too far from the subject of this address to enter here further upon the wonderful new field of research opened by the study of nuclear transmutations, which will be one of the main subjects of discussion among physicists at this meeting. The essential point for our argument is indeed not to be found in such new experience but in the obvious impossibility to account for common physical and chemical evidence on the basis of the well-established main features of Rutherford's atomic model without departing radically from the classical ideas of mechanics and electromagnetism. In fact, notwithstanding the insight provided by Newtonian mechanics into the harmony of planetary motions expressed by the Keplerian laws, the stability properties of mechanical models like the solar system which, when disturbed, have no tendency to return to their original state, have clearly no sufficient resemblance with the intrinsic stability of the electronic configurations of atoms that is responsible for the specific properties of the elements. Above all, this stability is strikingly illustrated by spectral analysis which, as is well known, has revealed that any element possesses a characteristic spectrum of sharp lines, independent of the external conditions to such an extent that it offers a means of identifying the material composition of even the most remote stars by spectroscopic observations.

A clue to the solution of this dilemma was, however, already provided by Planck's discovery of the elementary quantum of action which was the outcome of a very different line of physical research. As it is well known, Planck was led to this fundamental discovery by his ingenious analysis of such features of the thermal equilibrium between matter and radiation which, according to the general principles of thermodynamics, should be entirely independent of any specific properties of matter, and accordingly of any special ideas on atomic constitution. The existence of the elementary quantum of action expresses, in fact, a new trait of individuality of physical processes which is quite foreign to the classical laws of mechanics and electromagnetism and limits their validity essentially to those phenomena which involve actions large compared to the value of a single quantum, as given by Planck's new atomistic constant. This condition, though amply fulfilled in the phenomena of ordinary physical experience, does in no way hold for the behaviour of electrons in atoms, and it is indeed only the existence of the quantum of action which prevents the fusion of the electrons and the nucleus into a neutral massive corpuscle of practically infinitesimal extension.

The recognition of this situation suggested at once the description of the binding of each electron in the field around the nucleus as a succession of individual processes by which the atom is transferred from one of its so-called stationary states to another of these states, with emission of the released energy in the form of a single quantum of electromagnetic radiation. This view, intimately akin to Einstein's successful interpretation of the photoelectric effect, and borne out so convincingly by the beautiful researches of Franck and Hertz on the excitation of spectral lines by impacts of electrons on atoms, did in fact not only provide an immediate explanation of the puzzling general laws of line spectra disentangled by Balmer, Rydberg, and Ritz, but, with the help of spectroscopic evidence, led gradually to a systematic classification of the types of stationary binding of any electron in an atom, offering a complete explanation of the remarkable relationships between the physical and chemical properties of the elements, as expressed in the famous periodic table of Mendeleev. While such an interpretation of the properties of matter appeared as a realisation, even surpassing the dreams of the Pythagoreans, of the ancient ideal of reducing the formulation of the laws of nature to considerations of pure numbers, the basic assumption of the individuality of the atomic processes involved at the same time an essential renunciation of the detailed causal connection between physical events, which through the ages had been the unquestioned foundation of natural philosophy.

Not only was any question of a return to a mode of description consistent with the principle of causality excluded by unambiguous experience of the most varied kind, but it soon proved possible to develop the original primitive attempts at accounting for the existence of the quantum of action in atomic theory into a proper, essentially statistical atomic mechanics, fully comparable in consistency and completeness with the structure of classical mechanics of which it appears as a rational generalization. The establishment of this new so-called quantum mechanics which, as is well known, we owe above all to the ingenious contributions of the younger generation of physicists has, indeed, quite apart from its astounding fruitfulness in all branches of atomic physics and chemistry, essentially clarified the epistemological basis of the analysis and synthesis of atomic phenomena. The revision of the very problem of observation in this field, initiated by Heisenberg, one of the principal founders of quantum mechanics, has in fact led to the disclosure of hitherto disregarded presuppositions for the unambiguous use of even the most elementary concepts on which the description of natural phenomena rests. The critical point is here the recognition that any attempt to analyse, in the customary way of classical physics, the "individuality" of atomic processes, as conditioned by the quantum of action, will be frustrated by the unavoidable interaction between the atomic objects concerned and the measuring instruments indispensable for that purpose.

An immediate consequence of this situation is that observations regarding the behaviour of atomic objects obtained with different experimental arrangements cannot in general be combined in the usual way of classical physics. In particular, any imaginable procedure aiming at the coordination in space and time of the electrons in an atom will unavoidably involve an essentially uncontrollable exchange of momentum and energy between the atom and the measuring agencies, entirely annihilating the remarkable regularities of atomic stability for which the quantum of action is responsible. Conversely, any investigation of such regularities, the very account of which implies the conservation laws of energy and momentum, will in principle impose a renunciation as regards the space-time coordination of the individual electrons in the atom. Far from being inconsistent, the aspects of quantum phenomena revealed by experience obtained under such mutually exclusive conditions must thus be considered complementary in quite a novel way. The viewpoint of "complementarity" does, indeed, in no way mean an arbitrary renunciation as regards the analysis of atomic phenomena, but is on the contrary the expression of a rational synthesis of the wealth of experience in this field, which exceeds the limits to which the application of the concept of causality is naturally confined.

Notwithstanding the encouragement given to the pursuit of such inquiries by the great example of relativity theory which, just through the disclosure of unsuspected presuppositions for the unambiguous use of all physical concepts, opened new possibilities for the comprehension of apparently irreconcilable phenomena, we must realize that the situation met with in modern atomic theory is entirely unprecedented in the history of physical science. Indeed, the whole conceptual structure of classical physics, brought to so wonderful a unification and completion by Einstein's work, rests on the assumption, well adapted to our daily experience of physical phenomena, that it is possible to discriminate between the behaviour of material objects and the question of their observation. For a parallel to the lesson of atomic theory regarding the limited applicability of such customary idealisations, we must in fact turn to quite other branches of science, such as psychology, or even to that kind of epistemological problems with which already thinkers like Buddha and Lao Tse have been confronted, when trying to harmonize our position as spectators and actors in the great drama of existence. Still, the recognition of an analogy in the purely logical character of the problems which present themselves in so widely separated fields of human interest does in no way imply acceptance in atomic physics of any mysticism foreign to the true spirit of science, but on the contrary it gives us an incitation to examine whether the straightforward solution of the unexpected paradoxes met with in the application of our simplest concepts to atomic phenomena might not help us to clarify conceptual difficulties in other domains of experience.

There has also been no lack of suggestions to look for a direct correlation between life or free will and those features of atomic phenomena for the comprehension of which the frame of classical physics is obviously too narrow. In fact, it is possible to point out many characteristic features of the reactions of living organisms, like the sensitivity of visual perception or the induction of gene mutation by penetrating radiation, which undoubtedly involve an amplification of the effects of individual atomic processes, similar to that on which the experimental technique of atomic physics is essentially based. Still, the recognition that the fineness of organization and regulation mechanisms of living beings goes even so far beyond any previous expectation does in itself in no way enable us to account for the peculiar characteristics of life. Indeed, the so-called holistic and finalistic aspects of biological phenomena can certainly not be immediately explained by the feature of individuality of atomic processes disclosed by the discovery of the quantum of action; rather would the essentially statistical character of quantum mechanics at first sight seem even to increase the difficulties of understanding the proper biological regularities. In this dilemma, however, the general lesson of atomic theory suggests that the only way to reconcile the laws of physics with the concepts suited for a description of the phenomena of life is to examine the essential difference in the conditions of the observation of physical and biological phenomena.

First of all we must realize that every experimental arrangement with which we could study the behaviour of the atoms constituting an organism to the extent to which this can be done for single atoms in the fundamental experiments of atomic physics will exclude the possibility of maintaining the organism alive. The incessant exchange of matter which is inseparably connected with life will even imply the impossibility of regarding an organism as a well-defined system of material particles like the systems considered in any account of the ordinary physical and chemical properties of matter. In fact, we are led to conceive the proper biological regularities as representing laws of nature complementary to those appropriate to the account of the properties of inanimate bodies, in analogy with the complementary relationship between the stability properties of the atoms themselves and such behaviour of their constituent particles as allows of a description in terms of space-time coordination. In this sense, the existence of life itself should be considered, both as regards its definition and observation, as a basic postulate of biology, not susceptible of further analysis, in the same way as the existence of the quantum of action, together with the ultimate atomicity of matter, forms the elementary basis of atomic physics.

It will be seen that such a viewpoint is equally removed from the extreme doctrines of mechanism and vitalism. On the one hand, it condemns as irrelevant any comparison of living organisms with machines, be these the relatively simple constructions contemplated by the old iatro-physicists, or the most refined modern amplifier devices, the uncritical emphasis of which would expose us to deserve the nickname of "iatro-quantists." On the other hand, it rejects as irrational all such attempts at introducing some kind of special biological laws inconsistent with well-established physical and chemical regularities, as have in our days been revived under the impression of the wonderful revelations of embryology regarding cell growth and division. In this connection it must be especially remembered that the possibility of avoiding any such inconsistency within the frame of complementarity is given by the very fact that no result of biological investigation can be unambiguously described otherwise than in terms of physics and chemistry, just as any account of experience even in atomic physics must ultimately rest on the use of the concepts indispensable for a conscious recording of sense impressions.

The last remark brings us back into the realm of psychology, where the difficulties presented by the problems of definition and observation in scientific investigations have been clearly recognized long before such questions became acute in natural science. Indeed, the impossibility in psychical experience to distinguish between the phenomena themselves and their conscious perception clearly demands a renunciation of a simple causal description on the model of classical physics, and the very way in which words like "thoughts" and "feelings" are used to describe such experience reminds one most suggestively of the complementarity encountered in atomic physics.

I shall not here enter into any further detail but only emphasize that it is just this impossibility of distinguishing, in introspection, sharply between subject and object which provides the necessary latitude for the manifestation of volition. To connect free will more directly with limitation of causality in atomic physics, as it is often suggested, is, however, entirely foreign to the tendency underlying the remarks here made about biological problems.

In concluding this address I hope that the temerity of a physicist venturing so far outside his restricted domain of science may be forgiven in view of the most welcome opportunity of profitable discussion offered to physicists and biologists by this gathering to honour the memory of the great pioneer to whose fundamental discoveries both branches of science owe so much.

Natural Philosophy
and
Human Cultures

1938

It is only with great hesitation that I have accepted the kind invitation to address this assembly of distinguished representatives of the anthropological and ethnographical sciences of which I, as a physicist, have of course no first-hand knowledge. Still, on this special occasion when even the historical surroundings speak to every one of us about aspects of life other than those discussed at the regular congress proceedings, it might perhaps be of interest to try with a few words to draw your attention to the epistemological aspect of the latest development of natural philosophy and its bearing on general human problems. Notwithstanding the great separation between our different branches of knowledge, the new lesson which has been impressed upon physicists regarding the caution with which all usual conventions must be applied as soon as we are not concerned with everyday experience may, indeed, be suited to remind us in a novel way of the dangers, well known to humanists, of judging from our own standpoint cultures developed within other societies.

Of course it is impossible to distinguish sharply between natural philosophy and human culture. The physical sciences are, in fact, an integral part of our civilization, not only because our ever-increasing mastery of the forces of nature has so completely changed the material conditions of life, but also because the study of these sciences has contributed so much to clarify the background of our own existence. What has it not meant in this respect that we no more consider ourselves to be privileged in living at the centre of the universe, surrounded by less fortunate societies inhabiting the edges of the abyss, but, through the development of astronomy and geography, have realized that we all share a small spherical planet of the solar system which again is only a small part of still larger systems. How forceful an admonition about the relativity of all human judgments have we not also in our days received through the renewed revision of the presuppositions underlying the unambiguous use of even our most elementary concepts such as space and time, which, in disclosing the essential dependence of every physical phenomenon on the standpoint of the observer, has contributed so largely to the unity and beauty of our whole world-picture.

While the importance of these great achievements for our general outlook is commonly realized, it is hardly yet so as regards the unsuspected epistemological lesson which the opening of quite new realms of physical research has given us in the latest years. Our penetration into the world of atoms, hitherto closed to the eyes of man, is indeed an adventure which may be compared with the great journeys of discovery of the circumnavigators and the bold explorations of astronomers into the depths of celestial space. As is well known, the marvellous development of the art of physical experimentation not only has removed the last traces of the old belief that the coarseness of our senses would forever prevent us from obtaining direct information about individual atoms, but has even shown us that the atoms themselves consist of still smaller corpuscles which can be isolated and the properties of which can be investigated separately. At the same time we have, however, in this fascinating field of experience been taught that the laws of nature hitherto known, which constitute the grand edifice of classical physics, are valid only when we deal with bodies consisting of practically infinite numbers of atoms. The new knowledge concerning the behaviour of single atoms and atomic corpuscles has, in fact, revealed an unexpected limit for the subdivision of all physical actions extending far beyond the old doctrine of the limited divisibility of matter and giving every atomic process a peculiar individual character. This discovery has, in fact, yielded a quite new basis for the understanding of the intrinsic stability of atomic structures, which, in the last resort, conditions the regularities of all ordinary experience.

How radical a change in our attitude towards the description of nature this development of atomic physics has brought about is perhaps most clearly illustrated by the fact that even the principle of causality, so far regarded as the unquestioned foundation for all interpretation of natural phenomena, has proved too narrow a frame to embrace the peculiar regularities governing individual atomic processes. Certainly everyone will understand that physicists have needed very cogent reasons to renounce the ideal of causality itself; but in the study of atomic phenomena we have repeatedly been taught that questions which were believed to have received long ago their final answers had most unexpected surprises in store for us. You will surely all have heard about the riddles regarding the most elementary properties of light and matter which have puzzled physicists so much in recent years. The apparent contradictions which we have met in this respect are, in fact, as acute as those which gave rise to the development of the theory of relativity in the beginning of this century and have, just as the latter, only found their explanation by a closer examination of the limitation imposed by the new experiences themselves on the unambiguous use of the concepts entering into the description of the phenomena. While in relativity theory the decisive point was the recognition of the essentially different ways in which observers moving relatively to each other will describe the behaviour of given objects, the elucidation of the paradoxes of atomic physics has disclosed the fact that the unavoidable interaction between the objects and the measuring instruments sets an absolute limit to the possibility of speaking of a behaviour of atomic objects which is independent of the means of observation.

We are here faced with an epistemological problem quite new in natural philosophy, where all description of experiences has so far been based upon the assumption, already inherent in ordinary conventions of language, that it is possible to distinguish sharply between the behaviour of objects and the means of observation. This assumption is not only fully justified by all everyday experience but even constitutes the whole basis of classical physics, which, just through the theory of relativity, has received such a wonderful completion. As soon as we are dealing, however, with phenomena like individual atomic processes which, due to their very nature, are essentially determined by the interaction between the objects in question and the measuring instruments necessary for the definition of the experimental arrangements, we are, therefore, forced to examine more closely the question of what kind of knowledge can be obtained concerning the objects. In this respect we must, on the one hand, realize that the

aim of every physical experiment—to gain knowledge under repro-
ducible and communicable conditions—leaves us no choice but to use
everyday concepts, perhaps refined by the terminology of classical
physics, not only in all accounts of the construction and manipulation
of the measuring instruments but also in the description of the actual
experimental results. On the other hand, it is equally important to
understand that just this circumstance implies that no result of an
experiment concerning a phenomenon which, in principle, lies outside the range of classical physics can be interpreted as giving information about independent properties of the objects, but is inherently connected with a definite situation in the description of which
the measuring instruments interacting with the objects also enter
essentially. This last fact gives the straightforward explanation of
the apparent contradictions which appear when results about atomic
objects obtained by different experimental arrangements are tenta-
tively combined into a self-contained picture of the object.

Information regarding the behaviour of an atomic object obtained
under definite experimental conditions may, however, according to a
terminology often used in atomic physics, be adequately characterized as *complementary* to any information about the same object
obtained by some other experimental arrangement excluding the fulfilment of the first conditions. Although such kinds of information
cannot be combined into a single picture by means of ordinary concepts, they represent indeed equally essential aspects of any knowledge of the object in question which can be obtained in this domain.
The recognition of such a complementary character of the mechanical
analogies by which one has attempted to visualize the individual
radiative effects has, in fact, led to an entirely satisfactory solution
of the riddles of the properties of light alluded to above. In the same
way it is only by taking into consideration the complementary relationship between the different experiences concerning the behaviour
of atomic corpuscles that it has been possible to obtain a clue to the
understanding of the striking contrast between the properties of ordinary mechanical models and the peculiar laws of stability governing
atomic structures which form the basis for every closer explanation
of the specific physical and chemical properties of matter.

Of course I have no intention, on this occasion, of entering more
closely into such details, but I hope that I have been able to give you
a sufficiently clear impression of the fact that we are here in no way
concerned with an arbitrary renunciation as regards the detailed analysis of the almost overwhelming richness of our rapidly increasing
experience in the realm of atoms. On the contrary, we have to do

with a rational development of our means of classifying and comprehending new experience which, due to its very character, finds no place within the frame of causal description that is only suited to account for the behaviour of objects as long as this behaviour is independent of the means of observation. Far from containing any mysticism contrary to the spirit of science, the viewpoint of complementarity forms indeed a consistent generalization of the ideal of causality.

However unexpected this development may appear in the domain of physics, I am sure that many of you will have recognized the close analogy between the situation as regards the analysis of atomic phenomena, which I have described, and characteristic features of the problem of observation in human psychology. Indeed, we may say that the trend of modern psychology can be characterized as a reaction against the attempt at analyzing psychical experience into elements which can be associated in the same way as are the results of measurements in classical physics. In introspection it is clearly impossible to distinguish sharply between the phenomena themselves and their conscious perception, and although we may often speak of lending our attention to some particular aspect of a psychical experience, it will appear on closer examination that we really have to do, in such cases, with mutually exclusive situations. We all know the old saying that, if we try to analyze our own emotions, we hardly possess them any longer, and in that sense we recognize between psychical experiences, for the description of which words such as "thoughts" and "feelings" are adequately used, a complementary relationship similar to that between the experiences regarding the behaviour of atoms obtained under different experimental arrangements and described by means of different analogies taken from our usual ideas. By such a comparison it is, of course, in no way intended to suggest any closer relation between atomic physics and psychology, but merely to stress an epistemological argument common to both fields, and thus to encourage us to see how far the solution of the relatively simple physical problems may be helpful in clarifying the more intricate psychological questions with which human life confronts us, and which anthropologists and ethnologists so often meet in their investigations.

Coming now closer to our subject of the bearing of such viewpoints on the comparison of different human cultures, we shall first stress the typical complementary relationship between the modes of behaviour of living beings characterized by the words "instinct" and "reason." It is true that any such words are used in very different senses; thus, instinct may mean motive power or inherited behaviour, and reason may denote deeper sense as well as conscious argumentation. What we are concerned with is, however, only the practical way in which these words are used to discriminate between the different situations in which animals and men find themselves. Of course, nobody will deny our belonging to the animal world, and it would even be very difficult to find an exhaustive definition characterizing man among the other animals. Indeed, the latent possibilities in any living organism are not easily estimated, and I think that there is none of us who has not sometimes been deeply impressed by the extent to which circus animals can be drilled. Not even with respect to the conveyance of information from one individual to another would it be possible to draw a sharp separation between animals and man; but of course our power of speech places us in this respect in an essentially different situation, not only in the exchange of practical experience, but first of all in the possibility of transmitting to children, through education, the traditions concerning behaviour and reasoning which form the basis of any human culture.

As regards reason compared with instinct, it is, above all, essential to realize that no proper human thinking is imaginable without the use of concepts framed in some language which every generation has to learn anew. This use of concepts, in fact, not only is to a large extent suppressing instinctive life, but stands even largely in an exclusive relationship of complementarity to the display of inherited instincts. The astonishing superiority of lower animals compared with man in utilizing the possibilities of nature for the maintenance and propagation of life has certainly often its true explanation in the fact that on the part of such animals we can detect no conscious thinking, in our sense of the word. Similarly, the amazing capacity of so-called primitive people to orientate themselves in forests or deserts, which, though apparently lost in more civilized societies, may on occasion be revived in any of us, might justify the conclusion that such feats are only possible when no recourse is taken to conceptual thinking, which on its side is adapted to far more varied purposes of primary importance for the development of civilization. Just because it is not yet awake to the use of concepts, a newborn child can hardly be reckoned as a human being; but belonging to the species of man, it has, of course, though more helpless a creature than most young animals, the organic possibilities of receiving through education a culture which enables it to take its place in some human society.

Such considerations confront us at once with the question whether the widespread belief that every child is born with a predisposition for the adoption of a specific human culture is really well founded, or whether one has not rather to assume that any culture can be implanted and thrive on quite different physical backgrounds. Here we are of course touching a subject of still unsettled controversies between geneticists, who pursue most interesting studies on the inheritance of physical characters. In connection with such discussions, however, we must above all bear in mind that the distinction between the concepts genotype and phenotype, so fruitful for the clarification of heredity in plants and animals, essentially presupposes the subordinate influence of the external conditions of life on the characteristic properties of the species. In the case of the specific cultural characters of human societies the problem is, however, reversed in the sense that the basis for the classification is here the traditional habits shaped by the histories of the societies and their natural environments. These habits, as well as their inherent presuppositions, must therefore be analyzed in detail before any possible influence of inherited biological differences on the development and maintenance of the cultures concerned can be estimated. Indeed, in characterizing different nations and even different families within a nation, we may to a large extent consider biological traits and spiritual traditions as independent of each other, and it would even be tempting to reserve by definition the adjective "human" for those characters which are not directly bound to bodily inheritance.

At first sight, it might perhaps appear that such an attitude would mean unduly stressing merely dialectic points. But the lesson which we have received from the whole growth of the physical sciences is that the germ of fruitful development often lies just in the proper choice of definitions. When we think, for instance, of the clarification brought about in various branches of science by the argumentation of relativity theory, we see indeed what advance may lie in such formal refinements. As I have already hinted at earlier in this address, relativistic viewpoints are certainly also helpful in promoting a more objective attitude as to relationships between human cultures, the traditional differences of which in many ways resemble the different equivalent manners in which physical experience can be described. Still, this analogy between physical and humanistic problems is of limited scope and its exaggeration has even led to misunderstanding the essence of the theory of relativity itself. In fact, the unity of the relativistic world picture implies precisely the possibility for any one observer to predict within his own conceptual frame how any other observer will coordinate experience within the frame natural to him. The main obstacle to an unprejudiced attitude towards the relation between various human cultures is, however, the deep-rooted differences of the traditional backgrounds on which the cultural harmony in different human societies is based and which exclude any simple comparison between such cultures.

It is above all in this connection that the viewpoint of complementarity offers itself as a means of coping with the situation. In fact, when studying human cultures different from our own, we have to deal with a particular problem of observation which on closer consideration shows many features in common with atomic or psychological problems, where the interaction between objects and measuring tools, or the inseparability of objective content and observing subject, prevents an immediate application of the conventions suited to accounting for experiences of daily life. Especially in the study of cultures of primitive peoples, ethnologists not only are, indeed, aware of the risk of corrupting such cultures by the necessary contact, but are even confronted with the problem of the reaction of such studies on their own human attitude. What I here allude to is the experience, well known to explorers, of the shaking of their hitherto unrealized prejudices through the experience of the unsuspected inner harmony human life can present even under conventions and traditions most radically different from their own. As a specially drastic example I may perhaps here remind you of the extent to which in certain societies the roles of men and women are reversed, not only regarding domestic and social duties but also regarding behaviour and mentality. Even if many of us, in such a situation, might perhaps at first shrink from admitting the possibility that it is entirely a caprice of fate that the people concerned have their specific culture and not ours, and we not theirs instead of our own, it is clear that even the slightest suspicion in this respect implies a betrayal of the national complacency inherent in any human culture resting in itself.

Using the word much as it is used, in atomic physics, to characterize the relationship between experiences obtained by different experimental arrangements and visualizable only by mutually exclusive ideas, we may truly say that different human cultures are complementary to each other. Indeed, each such culture represents a harmonious balance of traditional conventions by means of which latent potentialities of human life can unfold themselves in a way which reveals to us new aspects of its unlimited richness and variety. Of course, there cannot, in this domain, be any question of such absolutely exclusive relationships as those between complementary experiences about the behaviour of well-defined atomic objects, since hardly any culture exists which could be said to be fully self-contained. On the contrary, we all know from numerous examples how a more or less intimate contact between different human societies can lead to a gradual fusion of traditions, giving birth to a quite new culture. The importance in this respect of the mixing of populations through emigration or conquest for the advancement of human civilization need hardly be recalled. It is, indeed, perhaps the greatest prospect of humanistic studies to contribute through an increasing knowledge of the history of cultural development to that gradual removal of prejudices which is the common aim of all science.

As I stressed in the beginning of this address, it is, of course, far beyond my capacities to contribute in any direct way to the solution of the problems discussed among the experts at this congress. My only purpose has been to give an impression of a general epistemological attitude which we have been forced to adopt in a field as far from human passions as the analysis of simple physical experiments. I do not know, however, whether I have found the right words to convey to you this impression, and before I conclude I may perhaps be allowed to relate an experience which once most vividly reminded me of my deficiencies in this respect. In order to explain to an audience that I did not use the word prejudice to imply any condemnation of other cultures, but merely to characterize our necessarily prejudiced conceptual frame, I referred jokingly to the traditional prejudices which the Danes cherish with regard to their Swedish brothers on the other side of the beautiful Sound outside these windows, with whom we have fought through centuries even within the walls of this castle, and from contact with whom we have, through the ages, received so much fruitful inspiration. Now you will realize what a shock I got when, after my address, a member of the audience came up to me and said that he could not understand why I hated the Swedes. Obviously I must have expressed myself most confusingly on that occasion, and I am afraid that also to-day I have talked in a very obscure way. Still, I hope that I have not spoken so unclearly as to give rise to any such misunderstandings of the trend of my argument.

Discussion with Einstein on Epistemological Problems in Atomic Physics

1949

When invited by the Editor of the series "Living Philosophers" to write an article for this volume in which contemporary scientists are honouring the epoch-making contributions of Albert Einstein to the progress of natural philosophy and are acknowledging the indebtedness of our whole generation for the guidance his genius has given us, I thought much of the best way of explaining how much I owe to him for inspiration. In this connection, the many occasions through the years on which I had the privilege to discuss with Einstein epistemological problems raised by the modern development of atomic physics have come back vividly to my mind and I have felt that I could hardly attempt anything better than to give an account of these discussions which have been of greatest value and stimulus to me. I hope also that the account may convey to wider circles an impression of how essential the open-minded exchange of ideas has been for the progress in a field where new experience has time after time demanded a reconsideration of our views.

From the very beginning the main point under debate has been the attitude to take to the departure from customary principles of natural philosophy characteristic of the novel development of physics which was initiated in the first year of this century by Planck's discovery of the universal quantum of action. This discovery, which revealed a feature of atomicity in the laws of nature going far beyond the old doctrine of the limited divisibility of matter, has indeed taught us that the classical theories of physics are idealizations which can be unambiguously applied only in the limit where all actions involved are large compared with the quantum. The question at issue has been whether the renunciation of a causal mode of description of atomic processes involved in the endeavours to cope with the situation should be regarded as a temporary departure from ideals to be ultimately revived or whether we are faced with an irrevocable step towards obtaining the proper harmony between analysis and synthesis of physical phenomena. To describe the background of our discussions and to bring out as clearly as possible the arguments for the contrasting viewpoints, I have felt it necessary to go to a certain length in recalling some main features of the development to which Einstein himself has contributed so decisively.

As is well known, it was the intimate relation, elucidated primarily by Boltzmann, between the laws of thermodynamics and the statistical regularities exhibited by mechanical systems with many degrees of freedom, which guided Planck in his ingenious treatment of the problem of thermal radiation, leading him to his fundamental discovery. While, in his work, Planck was principally concerned with considerations of essentially statistical character and with great caution refrained from definite conclusions as to the extent to which the existence of the quantum implied a departure from the foundations of mechanics and electrodynamics, Einstein's great original contribution to quantum theory (1905) was just the recognition of how physical phenomena like the photo-effect may depend directly on individual quantum effects.[1] In these very same years when, in the development of his theory of relativity, Einstein laid down a new foundation for physical science, he explored with a most daring spirit the novel features of atomicity which pointed beyond the framework of classical physics.

With unfailing intuition Einstein thus was led step by step to the conclusion that any radiation process involves the emission or absorption of individual light quanta or "photons" with energy and momentum

$$E = h\nu \quad \text{and} \quad P = h\sigma, \tag{1}$$

respectively, where h is Planck's constant, while ν and σ are the

[1] A. Einstein, *Ann. Phys.*, 17, 132 (1905).

number of vibrations per unit time and the number of waves per
unit length, respectively. Notwithstanding its fertility, the idea of
the photon implied a quite unforeseen dilemma, since any simple
corpuscular picture of radiation would obviously be irreconcilable
with interference effects, which present so essential an aspect of radiative phenomena, and which can be described only in terms of a wave
picture. The acuteness of the dilemma is stressed by the fact that
the interference effects offer our only means of defining the concepts
of frequency and wave-length entering into the very expressions
for the energy and momentum of the photon.

In this situation, there could be no question of attempting a causal
analysis of radiative phenomena, but only, by a combined use of
the contrasting pictures, to estimate probabilities for the occurrence
of the individual radiation processes. However, it is most important
to realize that the recourse to probability laws under such circumstances is essentially different in aim from the familiar application of
statistical considerations as practical means of accounting for the
properties of mechanical systems of great structural complexity. In
fact, in quantum physics we are presented not with intricacies of this
kind, but with the inability of the classical frame of concepts to com-
prise the peculiar feature of indivisibility, or "individuality," characterizing the elementary processes.

The failure of the theories of classical physics in accounting for
atomic phenomena was further accentuated by the progress of our
knowledge of the structure of atoms. Above all, Rutherford's discovery of the atomic nucleus (1911) revealed at once the inadequacy of classical mechanical and electromagnetic concepts to explain the inherent stability of the atom. Here again the quantum
theory offered a clue for the elucidation of the situation and especially
it was found possible to account for the atomic stability, as well as
for the empirical laws governing the spectra of the elements, by assuming that any reaction of the atom resulting in a change of its
energy involved a complete transition between two so-called stationary quantum states and that, in particular, the spectra were
emitted by a step-like process in which each transition is accompanied by the emission of a monochromatic light quantum of an
energy just equal to that of an Einstein photon.

These ideas, which were soon confirmed by the experiments of
Franck and Hertz (1914) on the excitation of spectra by impact of
electrons on atoms, involved a further renunciation of the causal
mode of description, since evidently the interpretation of the spectral
laws implies that an atom in an excited state in general will have the

possibility of transitions with photon emission to one or another of its lower energy states. In fact, the very idea of stationary states is incompatible with any directive for the choice between such transitions and leaves room only for the notion of the relative probabilities of the individual transition processes. The only guide in estimating such probabilities was the so-called correspondence principle which originated in the search for the closest possible connection between the statistical account of atomic processes and the consequences to be expected from classical theory, which should be valid in the limit where the actions involved in all stages of the analysis of the phenomena are large compared with the universal quantum.

At that time, no general self-consistent quantum theory was yet in sight, but the prevailing attitude may perhaps be illustrated by the following passage from a lecture by the writer from 1913: [2]

I hope that I have expressed myself sufficiently clearly so that you may appreciate the extent to which these considerations conflict with the admirably consistent scheme of conceptions which has been rightly termed the classical theory of electrodynamics. On the other hand, I have tried to convey to you the impression that—just by emphasizing so strongly this conflict—it may also be possible in course of time to establish a certain coherence in the new ideas.

Important progress in the development of quantum theory was made by Einstein himself in his famous article on radiative equilibrium in 1917,[3] where he showed that Planck's law for thermal radiation could be simply deduced from assumptions conforming with the basic ideas of the quantum theory of atomic constitution. To this purpose, Einstein formulated general statistical rules regarding the occurrence of radiative transitions between stationary states, assuming not only that, when the atom is exposed to a radiation field, absorption as well as emission processes will occur with a probability per unit time proportional to the intensity of the irradiation, but that even in the absence of external disturbances spontaneous emission processes will take place with a rate corresponding to a certain a priori probability. Regarding the latter point, Einstein emphasized the fundamental character of the statistical description in a most suggestive way by drawing attention to the analogy beween the assumptions regarding the occurrence of the spontaneous radiative transitions and the well-known laws governing transformations of radioactive substances.

[2] N. Bohr, The Theory of Spectra and Atomic Constitution, Cambridge, University Press, 1922.
[3] A. Einstein, Physik. Z., 18, 121 (1917).

In connection with a thorough examination of the exigencies of thermodynamics as regards radiation problems, Einstein stressed the dilemma still further by pointing out that the argumentation implied that any radiation process was "unidirected" in the sense that not only is a momentum corresponding to a photon with the direction of propagation transferred to an atom in the absorption process, but that also the emitting atom will receive an equivalent impulse in the opposite direction, although there can on the wave picture be no question of a preference for a single direction in an emission process. Einstein's own attitude to such startling conclusions is expressed in a passage at the end of the article (*loc. cit.*, p. 127f.), which may be translated as follows:

These features of the elementary processes would seem to make the development of a proper quantum treatment of radiation almost unavoidable. The weakness of the theory lies in the fact that, on the one hand, no closer connection with the wave concepts is obtainable and that, on the other hand, it leaves to chance (*Zufall*) the time and the direction of the elementary processes; nevertheless, I have full confidence in the reliability of the way entered upon.

When I had the great experience of meeting Einstein for the first time during a visit to Berlin in 1920, these fundamental questions formed the theme of our conversations. The discussions, to which I have often reverted in my thoughts, added to all my admiration for Einstein a deep impression of his detached attitude. Certainly, his favoured use of such picturesque phrases as "ghost waves (*Gespensterfelder*) guiding the photons" implied no tendency to mysticism, but illuminated rather a profound humour behind his piercing remarks. Yet, a certain difference in attitude and outlook remained, since, with his mastery for co-ordinating apparently contrasting experience without abandoning continuity and causality, Einstein was perhaps more reluctant to renounce such ideals than someone for whom renunciation in this respect appeared to be the only way open to proceed with the immediate task of co-ordinating the multifarious evidence regarding atomic phenomena, which accumulated from day to day in the exploration of this new field of knowledge.

In the following years, during which the atomic problems attracted the attention of rapidly increasing circles of physicists, the apparent contradictions inherent in quantum theory were felt ever more acutely. Illustrative of this situation is the discussion raised by the discovery of the Stern-Gerlach effect in 1922. On the one hand, this effect gave striking support to the idea of stationary states and in particular to the quantum theory of the Zeeman effect developed by Sommerfeld; on the other hand, as exposed so clearly by Einstein and Ehrenfest,[4] it presented with unsurmountable difficulties any attempt at forming a picture of the behaviour of atoms in a magnetic field. Similar paradoxes were raised by the discovery by Compton (1924) of the change in wave-length accompanying the scattering of X-rays by electrons. This phenomenon afforded, as is well known, a most direct proof of the adequacy of Einstein's view regarding the transfer of energy and momentum in radiative processes; at the same time, it was equally clear that no simple picture of a corpuscular collision could offer an exhaustive description of the phenomenon. Under the impact of such difficulties, doubts were for a time entertained even regarding the conservation of energy and momentum in the individual radiation processes;[5] a view, however, which very soon had to be abandoned in face of more refined experiments bringing out the correlation between the deflection of the photon and the corresponding electron recoil.

The way to the clarification of the situation was, indeed, first to be paved by the development of a more comprehensive quantum theory. A first step towards this goal was the recognition by de Broglie in 1925 that the wave-corpuscle duality was not confined to the properties of radiation, but was equally unavoidable in accounting for the behaviour of material particles. This idea, which was soon convincingly confirmed by experiments on electron interference phenomena, was at once greeted by Einstein, who had already envisaged the deep-going analogy between the properties of thermal radiation and of gases in the so-called degenerate state.[6] The new line was pursued with the greatest success by Schrödinger (1926) who, in particular, showed how the stationary states of atomic systems could be represented by the proper solutions of a wave-equation to the establishment of which he was led by the formal analogy, originally traced by Hamilton, between mechanical and optical problems. Still, the paradoxical aspects of quantum theory were in no way ameliorated, but even emphasized, by the apparent contradiction between the exigencies of the general superposition principle of the wave description and the feature of individuality of the elementary atomic processes.

[4] A. Einstein and P. Ehrenfest, Z. *Physik*, *11*, 31 (1922).
[5] N. Bohr, H. A. Kramers and J. C. Slater, *Phil. Mag.*, *47*, 785 (1924).
[6] A. Einstein, *Berl. Ber.* 261 (1924); 3 and 18 (1925).

At the same time, Heisenberg (1925) had laid the foundation of a rational quantum mechanics, which was rapidly developed through important contributions by Born and Jordan as well as by Dirac. In this theory, a formalism is introduced, in which the kinematical and dynamical variables of classical mechanics are replaced by symbols subjected to a non-commutative algebra. Notwithstanding the re-nunciation of orbital pictures, Hamilton's canonical equations of me-chanics are kept unaltered and Planck's constant enters only in the rules of commutation

$$qp - pq = \sqrt{-1}\,\frac{h}{2\pi} \tag{2}$$

holding for any set of conjugate variables q and p. Through a representation of the symbols by matrices with elements referring to transitions between stationary states, a quantitative formulation of the correspondence principle became for the first time possible. It may here be recalled that an important preliminary step towards this goal was reached through the establishment, especially by con-tributions of Kramers, of a quantum theory of dispersion making basic use of Einstein's general rules for the probability of the occur-rence of absorption and emission processes.

This formalism of quantum mechanics was soon proved by Schrö-dinger to give results identical with those obtainable by the mathe-matically often more convenient methods of wave theory, and in the following years general methods were gradually established for an essentially statistical description of atomic processes combining the features of individuality and the requirements of the superposi-tion principle, equally characteristic of quantum theory. Among the many advances in this period, it may especially be mentioned that the formalism proved capable of incorporating the exclusion principle which governs the states of systems with several electrons, and which already before the advent of quantum mechanics had been derived by Pauli from an analysis of atomic spectra. The quantitative com-prehension of a vast amount of empirical evidence could leave no doubt as to the fertility and adequacy of the quantum-mechanical formalism, but its abstract character gave rise to a widespread feel-ing of uneasiness. An elucidation of the situation should, indeed, demand a thorough examination of the very observational problem in atomic physics.

This phase of the development was, as is well known, initiated in 1927 by Heisenberg,[1] who pointed out that the knowledge obtainable

[1] W. Heisenberg, Z. *Physik*, *43*, 172 (1927).

of the state of an atomic system will always involve a peculiar "indeterminacy." Thus, any measurement of the position of an electron by means of some device, like a microscope, making use of high-frequency radiation, will, according to the fundamental relations (1), be connected with a momentum exchange between the electron and the measuring agency, which is the greater the more accurate a position measurement is attempted. In comparing such considerations with the exigencies of the quantum-mechanical formalism, Heisenberg called attention to the fact that the commutation rule (2) imposes a reciprocal limitation on the fixation of two conjugate variables, q and p, expressed by the relation

$$\Delta q \cdot \Delta p \approx h, \tag{3}$$

where Δq and Δp are suitably defined latitudes in the determination of these variables. In pointing to the intimate connection between the statistical description in quantum mechanics and the actual possibilities of measurement, this so-called indeterminacy relation is, as Heisenberg showed, most important for the elucidation of the paradoxes involved in the attempts of analyzing quantum effects with reference to customary physical pictures.

The new progress in atomic physics was commented upon from various sides at the International Physical Congress held in September 1927 at Como in commemoration of Volta. In a lecture on that occasion,[8] I advocated a point of view conveniently termed "complementarity," suited to embrace the characteristic features of individuality of quantum phenomena, and at the same time to clarify the peculiar aspects of the observational problem in this field of experience. For this purpose, it is decisive to recognize that, *however far the phenomena transcend the scope of classical physical explanation, the account of all evidence must be expressed in classical terms.* The argument is simply that by the word "experiment" we refer to a situation where we can tell others what we have done and what we have learned and that, therefore, the account of the experimental arrangement and of the results of the observations must be expressed in unambiguous language with suitable application of the terminology of classical physics.

This crucial point, which was to become a main theme of the discussions reported in the following, implies the *impossibility of any sharp separation between the behaviour of atomic objects and the interaction with the measuring instruments which serve to define the*

[8] Atti del Congresso Internazionale dei Fisici, Como, Settembre 1927 (reprinted in *Nature, 121,* 78 and 580, 1928).

conditions under which the phenomena appear. In fact, the individuality of the typical quantum effects finds its proper expression in the circumstance that any attempt of subdividing the phenomena will demand a change in the experimental arrangement introducing new possibilities of interaction between objects and measuring instruments which in principle cannot be controlled. Consequently, evidence obtained under different experimental conditions cannot be comprehended within a single picture, but must be regarded as *complementary* in the sense that only the totality of the phenomena exhausts the possible information about the objects.

Under these circumstances an essential element of ambiguity is involved in ascribing conventional physical attributes to atomic objects, as is at once evident in the dilemma regarding the corpuscular and wave properties of electrons and photons, where we have to do with contrasting pictures, each referring to an essential aspect of empirical evidence. An illustrative example, of how the apparent paradoxes are removed by an examination of the experimental conditions under which the complementary phenomena appear, is also given by the Compton effect, the consistent description of which at first had presented us with such acute difficulties. Thus, any arrangement suited to study the exchange of energy and momentum between the electron and the photon must involve a latitude in the space-time description of the interaction sufficient for the definition of wave-number and frequency which enter into the relation (1). Conversely, any attempt of locating the collision between the photon and the electron more accurately would, on account of the unavoidable interaction with the fixed scales and clocks defining the space-time reference frame, exclude all closer account as regards the balance of momentum and energy.

As stressed in the lecture, an adequate tool for a complementary way of description is offered precisely by the quantum-mechanical formalism which represents a purely symbolic scheme permitting only predictions, on lines of the correspondence principle, as to results obtainable under conditions specified by means of classical concepts. It must here be remembered that even in the indeterminacy relation (3) we are dealing with an implication of the formalism which defies unambiguous expression in words suited to describe classical physical pictures. Thus, a sentence like "we cannot know both the momentum and the position of an atomic object" raises at once questions as to the physical reality of two such attributes of the object, which can be answered only by referring to the conditions for the unambiguous use of space-time concepts, on the one hand, and dynamical conservation laws, on the other hand. While the combination of these concepts into a single picture of a causal chain of events is the essence of classical mechanics, room for regularities beyond the grasp of such a description is just afforded by the circumstance that the study of the complementary phenomena demands mutually exclusive experimental arrangements.

The necessity, in atomic physics, of a renewed examination of the foundation for the unambiguous use of elementary physical ideas recalls in some way the situation that led Einstein to his original revision of the basis for all application of space-time concepts which, by its emphasis on the primordial importance of the observational problem, has lent such unity to our world picture. Notwithstanding all novelty of approach, causal description is upheld in relativity theory within any given frame of reference, but in quantum theory the uncontrollable interaction between the objects and the measuring instruments forces us to a renunciation even in such respect. This recognition, however, in no way points to any limitation of the scope of the quantum-mechanical description, and the trend of the whole argumentation presented in the Como lecture was to show that the viewpoint of complementarity may be regarded as a rational generalization of the very ideal of causality.

At the general discussion in Como, we all missed the presence of Einstein, but soon after, in October 1927, I had the opportunity to meet him in Brussels at the Fifth Physical Conference of the Solvay Institute, which was devoted to the theme "Electrons and Photons." At the Solvay meetings, Einstein had from their beginning been a most prominent figure, and several of us came to the conference with great anticipations to learn his reaction to the latest stage of the development which, to our view, went far in clarifying the problems which he had himself from the outset elicited so ingeniously. During the discussions, where the whole subject was reviewed by contributions from many sides and where also the arguments mentioned in the preceding pages were again presented, Einstein expressed, however, a deep concern over the extent to which causal account in space and time was abandoned in quantum mechanics.

To illustrate his attitude, Einstein referred at one of the sessions [9]

[9] Institut International de Physique Solvay, *Rapport et discussions* du 5ᵉ Conseil, Paris 1928, 253ff.

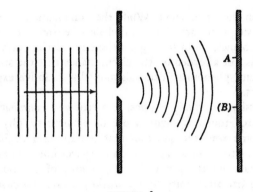

FIGURE 1

to the simple example, illustrated by Figure 1, of a particle (electron or photon) penetrating through a hole or a narrow slit in a diaphragm placed at some distance before a photographic plate. On account of the diffraction of the wave connected with the motion of the particle and indicated in the figure by the thin lines, it is under such conditions not possible to predict with certainty at what point the electron will arrive at the photographic plate, but only to calculate the probability that, in an experiment, the electron will be found within any given region of the plate. The apparent difficulty, in this description, which Einstein felt so acutely, is the fact that, if in the experiment the electron is recorded at one point A of the plate, then it is out of the question of ever observing an effect of this electron at another point (B), although the laws of ordinary wave propagation offer no room for a correlation between two such events.

Einstein's attitude gave rise to ardent discussions within a small circle, in which Ehrenfest, who through the years had been a close friend of us both, took part in a most active and helpful way. Surely, we all recognized that, in the above example, the situation presents no analogue to the application of statistics in dealing with complicated mechanical systems, but rather recalled the background for Einstein's own early conclusions about the unidirection of individual radiation effects which contrasts so strongly with a simple wave picture (cf. p. 36). The discussions, however, centered on the question of whether the quantum-mechanical description exhausted the possibilities of accounting for observable phenomena or, as Einstein maintained, the analysis could be carried further and, especially, of whether a fuller description of the phenomena could be obtained by bringing into consideration the detailed balance of energy and momentum in individual processes.

To explain the trend of Einstein's arguments, it may be illustrative here to consider some simple features of the momentum and energy balance in connection with the location of a particle in space and time. For this purpose, we shall examine the simple case of a particle penetrating through a hole in a diaphragm without or with a shutter to open and close the hole, as indicated in Figures 2a and 2b, respectively. The equidistant parallel lines to the left in the figures indicate the train of plane waves corresponding to the state of motion of a particle which, before reaching the diaphragm, has a momentum P related to the wave-number σ by the second of equations (1). In accordance with the diffraction of the waves when passing through the hole, the state of motion of the particle to the right of the diaphragm is represented by a spherical wave train with a suitably defined angular aperture θ and, in case of Figure 2b, also with a limited radial extension. Consequently, the description of this state involves a certain latitude Δp in the momentum component of the particle parallel to the diaphragm and, in the case of a diaphragm with a shutter, an additional latitude ΔE of the kinetic energy.

Since a measure for the latitude Δq in location of the particle in the plane of the diaphragm is given by the radius a of the hole, and since $\theta \approx 1/\sigma a$, we get, using (1), just $\Delta p \approx \theta P \approx h/\Delta q$, in accordance with the indeterminacy relation (3). This result could, of course, also be obtained directly by noticing that, due to the limited extension of the wave-field at the place of the slit, the component of the wave-number parallel to the plane of the diaphragm will involve a latitude $\Delta \sigma \approx 1/a \approx 1/\Delta q$. Similarly, the spread of the fre-

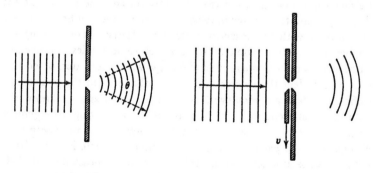

FIGURE 2a   FIGURE 2b

quencies of the harmonic components in the limited wave-train in
Figure 2b is evidently $\Delta v \approx 1/\Delta t$, where Δt is the time interval during which the shutter leaves the hole open and, thus, represents
the latitude in time of the passage of the particle through the diaphragm. From (1), we therefore get

$$\Delta E \cdot \Delta t \approx h, \tag{4}$$

again in accordance with the relation (3) for the two conjugated
variables E and t.

From the point of view of the laws of conservation, the origin of
such latitudes entering into the description of the state of the particle
after passing through the hole may be traced to the possibilities of
momentum and energy exchange with the diaphragm or the shutter.
In the reference system considered in Figures 2a and 2b, the velocity
of the diaphragm may be disregarded and only a change of momentum Δp between the particle and the diaphragm needs to be taken
into consideration. The shutter, however, which leaves the hole
opened during the time Δt, moves with a considerable velocity $v \approx
a/\Delta t$, and a momentum transfer Δp involves therefore an energy ex-
change with the particle, amounting to

$$v \, \Delta p \approx \frac{\Delta q \, \Delta p}{\Delta t} \approx \frac{h}{\Delta t},$$

being just of the same order of magnitude as the latitude ΔE given by
(4) and, thus, allowing for momentum and energy balance.

The problem raised by Einstein was now to what extent a control
of the momentum and energy transfer, involved in a location of the
particle in space and time, can be used for a further specification of
the state of the particle after passing through the hole. Here, it must
be taken into consideration that the position and the motion of the
diaphragm and the shutter have so far been assumed to be accurately
coordinated with the space-time reference frame. This assumption
implies, in the description of the state of these bodies, an essential
latitude as to their momentum and energy which need not, of course,
noticeably affect the velocities, if the diaphragm and the shutter are
sufficiently heavy. However, as soon as we want to know the
momentum and energy of these parts of the measuring arrangement
with an accuracy sufficient to control the momentum and energy exchange with the particle under investigation, we shall, in accordance
with the general indeterminacy relations, lose the possibility of their
accurate location in space and time. We have, therefore, to examine

how far this circumstance will affect the intended use of the whole arrangement and, as we shall see, this crucial point clearly brings out the complementary character of the phenomena.

Returning for a moment to the case of the simple arrangement indicated in Figure 1, it has so far not been specified to what use it is intended. In fact, it is only on the assumption that the diaphragm and the plate have well-defined positions in space that it is impossible, within the frame of the quantum-mechanical formalism, to make more detailed predictions as to the point of the photographic plate where the particle will be recorded. If, however, we admit a sufficiently large latitude in the knowledge of the position of the diaphragm, it should, in principle, be possible to control the momentum transfer to the diaphragm and, thus, to make more detailed predictions as to the direction of the electron path from the hole to the recording point. As regards the quantum-mechanical description, we have to deal here with a two-body system consisting of the diaphragm as well as of the particle, and it is just with an explicit application of conservation laws to such a system that we are concerned in the Compton effect where, for instance, the observation of the recoil of the electron by means of a cloud chamber allows us to predict in what direction the scattered photon will eventually be observed.

The importance of considerations of this kind was, in the course of the discussions, most interestingly illuminated by the examination of an arrangement where between the diaphragm with the slit and the photographic plate is inserted another diaphragm with two parallel slits, as is shown in Figure 3. If a parallel beam of electrons (or photons) falls from the left on the first diaphragm, we shall, under usual conditions, observe on the plate an interference pattern indicated by the shading of the photographic plate shown in front view to the right of the figure. With intense beams, this pattern is built up by

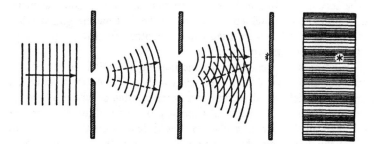

FIGURE 3

the accumulation of a large number of individual processes, each giving rise to a small spot on the photographic plate, and the distribution of these spots follows a simple law derivable from the wave analysis. The same distribution should also be found in the statistical account of many experiments performed with beams so faint that in a single exposure only one electron (or photon) will arrive at the photographic plate at some spot shown in the figure as a small star. Since, now, as indicated by the broken arrows, the momentum transferred to the first diaphragm ought to be different if the electron was assumed to pass through the upper or the lower slit in the second diaphragm, Einstein suggested that a control of the momentum transfer would permit a closer analysis of the phenomenon and, in particular, make it possible to decide through which of the two slits the electron had passed before arriving at the plate.

A closer examination showed, however, that the suggested control of the momentum transfer would involve a latitude in the knowledge of the position of the diaphragm which would exclude the appearance of the interference phenomena in question. In fact, if ω is the small angle between the conjectured paths of a particle passing through the upper or the lower slit, the difference of momentum transfer in these two cases will, according to (1), be equal to $h\sigma\omega$, and any control of the momentum of the diaphragm with an accuracy sufficient to measure this difference will, due to the indeterminacy relation, involve a minimum latitude of the position of the diaphragm, comparable with $1/\sigma\omega$. If, as in the figure, the diaphragm with the two slits is placed in the middle between the first diaphragm and the photographic plate, it will be seen that the number of fringes per unit length will be just equal to $\sigma\omega$ and, since an uncertainty in the position of the first diaphragm of the amount of $1/\sigma\omega$ will cause an equal uncertainty in the positions of the fringes, it follows that no interference effect can appear. The same result is easily shown to hold for any other placing of the second diaphragm between the first diaphragm and the plate, and would also be obtained if, instead of the first diaphragm, another of these three bodies were used for the control, for the purpose suggested, of the momentum transfer.

This point is of great logical consequence, since it is only the circumstance that we are presented with a choice of *either* tracing the path of a particle *or* observing interference effects, which allows us to escape from the paradoxical necessity of concluding that the behaviour of an electron or a photon should depend on the presence of a slit in the diaphragm through which it could be proved not to pass. We have here to do with a typical example of how the complementary phenomena appear under mutually exclusive experimental arrangements (cf. p. 40) and are just faced with the impossibility, in the analysis of quantum effects, of drawing any sharp separation between an independent behaviour of atomic objects and their interaction with the measuring instruments which serve to define the conditions under which the phenomena occur.

Our talks about the attitude to be taken in face of a novel situation as regards analysis and synthesis of experience touched naturally on many aspects of philosophical thinking, but, in spite of all divergencies of approach and opinion, a most humorous spirit animated the discussions. On his side, Einstein mockingly asked us whether we could really believe that the providential authorities took recourse to dice-playing (". . . ob der liebe Gott würfelt"), to which I replied by pointing at the great caution, already called for by ancient thinkers, in ascribing attributes to Providence in everyday language. I remember also how at the peak of the discussion Ehrenfest, in his affectionate manner of teasing his friends, jokingly hinted at the apparent similarity between Einstein's attitude and that of the opponents of relativity theory; but instantly Ehrenfest added that he would not be able to find relief in his own mind before concord with Einstein was reached.

Einstein's concern and criticism provided a most valuable incentive for us all to reexamine the various aspects of the situation as regards the description of atomic phenomena. To me it was a welcome stimulus to clarify still further the rôle played by the measuring instruments and, in order to bring into strong relief the mutually exclusive character of the experimental conditions under which the complementary phenomena appear, I tried in those days to sketch various apparatus in a pseudo-realistic style of which the following figures are examples. Thus, for the study of an interference phenomenon of the type indicated in Figure 3, it suggests itself to use an experimental arrangement like that shown in Figure 4, where the solid parts of the apparatus, serving as diaphragms and plate-holder, are firmly bolted to a common support. In such an arrangement, where the knowledge of the relative positions of the diaphragms and the photographic plate is secured by a rigid connection, it is obviously impossible to control the momentum exchanged between the particle and the separate parts of the apparatus. The only way in which, in such an arrangement, we could insure that the particle passed through one of the slits in the second diaphragm is to cover the other slit by

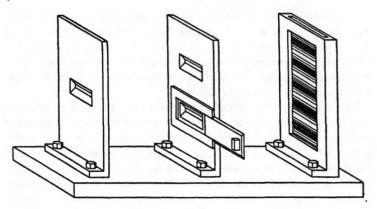

FIGURE 4

a lid, as indicated in the figure; but if the slit is covered, there is of course no question of any interference phenomenon, and on the plate we shall simply observe a continuous distribution as in the case of the single fixed diaphragm in Figure 1.

In the study of phenomena in the account of which we are dealing with detailed momentum balance, certain parts of the whole device

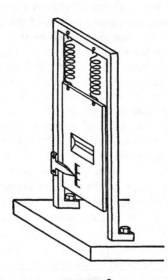

FIGURE 5

must naturally be given the freedom to move independently of others. Such an apparatus is sketched in Figure 5, where a diaphragm with a slit is suspended by weak springs from a solid yoke bolted to the support on which also other immobile parts of the arrangement are to be fastened. The scale on the diaphragm together with the pointer on the bearings of the yoke refer to such study of the motion of the diaphragm, as may be required for an estimate of the momentum transferred to it, permitting one to draw conclusions as to the deflection suffered by the particle in passing through the slit. Since, however, any reading of the scale, in whatever way performed, will involve an uncontrollable change in the momentum of the diaphragm, there will always be, in conformity with the indeterminacy principle, a reciprocal relationship between our knowledge of the position of the slit and the accuracy of the momentum control.

In the same semi-serious style, Figure 6 represents a part of an arrangement suited for the study of phenomena which, in contrast to those just discussed, involve time coordination explicitly. It consists of a shutter rigidly connected with a robust clock resting on the support which carries a diaphragm and on which further parts of similar character, regulated by the same clock-work or by other clocks standardized relatively to it, are also to be fixed. The special aim of the figure is to underline that a clock is a piece of machinery, the working of which can completely be accounted for by ordinary

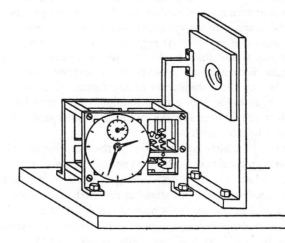

FIGURE 6

mechanics and will be affected neither by reading of the position of its hands nor by the interaction between its accessories and an atomic particle. In securing the opening of the hole at a definite moment, an apparatus of this type might, for instance, be used for an accurate measurement of the time an electron or a photon takes to come from the diaphragm to some other place, but, evidently, it would leave no possibility of controlling the energy transfer to the shutter with the aim of drawing conclusions as to the energy of the particle which has passed through the diaphragm. If we are interested in such conclusions we must, of course, use an arrangement where the shutter devices can no longer serve as accurate clocks, but where the knowledge of the moment when the hole in the diaphragm is open involves a latitude connected with the accuracy of the energy measurement by the general relation (4).

The contemplation of such more or less practical arrangements and their more or less fictitious use proved most instructive in directing attention to essential features of the problems. The main point here is the distinction between the *objects* under investigation and the *measuring instruments* which serve to define, in classical terms, the conditions under which the phenomena appear. Incidentally, we may remark that, for the illustration of the preceding considerations, it is not relevant that experiments involving an accurate control of the momentum or energy transfer from atomic particles to heavy bodies like diaphragms and shutters would be very difficult to perform, if practicable at all. It is only decisive that, in contrast to the proper measuring instruments, these bodies together with the particles would in such a case constitute the system to which the quantum-mechanical formalism has to be applied. As regards the specification of the conditions for any well-defined application of the formalism, it is moreover essential that the *whole experimental arrangement* be taken into account. In fact, the introduction of any further piece of apparatus, like a mirror, in the way of a particle might imply new interference effects essentially influencing the predictions as regards the results to be eventually recorded.

The extent to which renunciation of the visualization of atomic phenomena is imposed upon us by the impossibility of their subdivision is strikingly illustrated by the following example to which Einstein very early called attention and often has reverted. If a semi-reflecting mirror is placed in the way of a photon, leaving two possibilities for its direction of propagation, the photon may either be recorded on one, and only one, of two photographic plates situated at great distances in the two directions in question, or else we may, by replacing the plates by mirrors, observe effects exhibiting an interference between the two reflected wave-trains. In any attempt of a pictorial representation of the behaviour of the photon we would, thus, meet with the difficulty: to be obliged to say, on the one hand, that the photon always chooses *one* of the two ways and, on the other hand, that it behaves as if it had passed *both* ways.

It is just arguments of this kind which recall the impossibility of subdividing quantum phenomena and reveal the ambiguity in ascribing customary physical attributes to atomic objects. In particular, it must be realized that—besides in the account of the placing and timing of the instruments forming the experimental arrangement—all unambiguous use of space-time concepts in the description of atomic phenomena is confined to the recording of observations which refer to marks on a photographic plate or to similar practically irreversible amplification effects like the building of a water drop around an ion in a cloud-chamber. Although, of course, the existence of the quantum of action is ultimately responsible for the properties of the materials of which the measuring instruments are built and on which the functioning of the recording devices depends, this circumstance is not relevant for the problems of the adequacy and completeness of the quantum-mechanical description in its aspects here discussed.

These problems were instructively commented upon from different sides at the Solvay meeting,[10] in the same session where Einstein raised his general objections. On that occasion an interesting discussion arose also about how to speak of the appearance of phenomena for which only predictions of statistical character can be made. The question was whether, as to the occurrence of individual effects, we should adopt a terminology proposed by Dirac, that we were concerned with a choice on the part of "nature," or, as suggested by Heisenberg, we should say that we have to do with a choice on the part of the "observer" constructing the measuring instruments and reading their recording. Any such terminology would, however, appear dubious since, on the one hand, it is hardly reasonable to endow nature with volition in the ordinary sense, while, on the other hand, it is certainly not possible for the observer to influence the events which may appear under the conditions he has arranged. To my mind, there is no other alternative than to admit that, in this field of experience, we are dealing with individual phenomena and that our possibilities of handling the measuring instruments allow us only to make a choice between the different complementary types of phenomena we want to study.

[10] *Ibid.*, 248ff.

The epistemological problems touched upon here were more explicitly dealt with in my contribution to the issue of *Naturwissenschaften* in celebration of Planck's 70th birthday in 1929. In this article, a comparison was also made between the lesson derived from the discovery of the universal quantum of action and the development which has followed the discovery of the finite velocity of light and which, through Einstein's pioneer work, has so greatly clarified basic principles of natural philosophy. In relativity theory, the emphasis on the dependence of all phenomena on the reference frame opened quite new ways of tracing general physical laws of unparalleled scope. In quantum theory, it was argued, the logical comprehension of hitherto unsuspected fundamental regularities governing atomic phenomena has demanded the recognition that no sharp separation can be made between an independent behaviour of the objects and their interaction with the measuring instruments which define the reference frame.

In this respect, quantum theory presents us with a novel situation in physical science, but attention was called to the very close analogy with the situation as regards analysis and synthesis of experience, which we meet in many other fields of human knowledge and interest. As is well known, many of the difficulties in psychology originate in the different placing of the separation lines between object and subject in the analysis of various aspects of psychical experience. Actually, words like "thoughts" and "sentiments," equally indispensable to illustrate the variety and scope of conscious life, are used in a similar complementary way as are space-time coordination and dynamical conservation laws in atomic physics. A precise formulation of such analogies involves, of course, intricacies of terminology, and the writer's position is perhaps best indicated in a passage in the article, hinting at the mutually exclusive relationship which will always exist between the practical use of any word and attempts at its strict definition. The principal aim, however, of these considerations, which were not least inspired by the hope of influencing Einstein's attitude, was to point to perspectives of bringing general epistemological problems into relief by means of a lesson derived from the study of new, but fundamentally simple, physical experience.

At the next meeting with Einstein at the Solvay Conference in 1930, our discussions took quite a dramatic turn. As an objection to the view that a control of the interchange of momentum and energy between the objects and the measuring instruments was excluded if these instruments should serve their purpose of defining the space-time frame of the phenomena, Einstein brought forward the argument that such control should be possible when the exigencies of relativity theory were taken into consideration. In particular, the general relationship between energy and mass, expressed in Einstein's famous formula

$$E = mc^2, \tag{5}$$

should allow, by means of simple weighing, to measure the total energy of any system and, thus, in principle to control the energy transferred to it when it interacts with an atomic object.

As an arrangement suited for such purpose, Einstein proposed the device indicated in Figure 7, consisting of a box with a hole in its side, which could be opened or closed by a shutter moved by means of a clock-work within the box. If, in the beginning, the box contained a certain amount of radiation and the clock was set to open the shutter for a very short interval at a chosen time, it could be achieved that a single photon was released through the hole at a moment known with as great accuracy as desired. Moreover, it would apparently also be possible, by weighing the whole box before and after this event, to measure the energy of the photon with any accuracy wanted, in definite contradiction to the reciprocal indeterminacy of time and energy quantities in quantum mechanics.

This argument amounted to a serious challenge and gave rise to a thorough examination of the whole problem. At the outcome of the discussion, to which Einstein himself contributed effectively, it became clear, however, that the argument could not be upheld. In

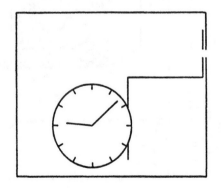

FIGURE 7

fact, in the consideration of the problem, it was found necessary to look closer into the consequences of the identification of inertial and gravitational mass implied in the application of relation (5). Especially, it was essential to take into account the relationship between the rate of a clock and its position in a gravitational field—well known from the red-shift of the lines in the sun's spectrum—following from Einstein's principle of equivalence between gravity effects and the phenomena observed in accelerated reference frames.

Our discussion concentrated on the possible application of an apparatus incorporating Einstein's device and drawn in Figure 8 in the same pseudo-realistic style as some of the preceding figures. The box, of which a section is shown in order to exhibit its interior, is suspended

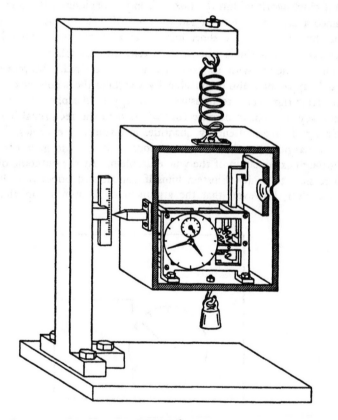

FIGURE 8

in a spring-balance and is furnished with a pointer to read its position on a scale fixed to the balance support. The weighing of the box may thus be performed with any given accuracy Δm by adjusting the balance to its zero position by means of suitable loads. The essential point is now that any determination of this position with a given accuracy Δq will involve a minimum latitude Δp in the control of the momentum of the box connected with Δq by the relation (3). This latitude must obviously again be smaller than the total impulse which, during the whole interval T of the balancing procedure, can be given by the gravitational field to a body with a mass Δm, or

$$\Delta p \approx \frac{h}{\Delta q} < T \cdot g \cdot \Delta m, \qquad (6)$$

where g is the gravity constant. The greater the accuracy of the reading q of the pointer, the longer must, consequently, be the balancing interval T, if a given accuracy Δm of the weighing of the box with its content shall be obtained.

Now, according to general relativity theory, a clock, when displaced in the direction of the gravitational force by an amount of Δq, will change its rate in such a way that its reading in the course of a time interval T will differ by an amount ΔT given by the relation

$$\frac{\Delta T}{T} = \frac{1}{c^2} g \, \Delta q. \qquad (7)$$

By comparing (6) and (7) we see, therefore, that after the weighing procedure there will in our knowledge of the adjustment of the clock be a latitude

$$\Delta T > \frac{h}{c^2 \, \Delta m}.$$

Together with the formula (5), this relation again leads to

$$\Delta T \cdot \Delta E > h,$$

in accordance with the indeterminacy principle. Consequently, a use of the apparatus as a means of accurately measuring the energy of the photon will prevent us from controlling the moment of its escape.

The discussion, so illustrative of the power and consistency of relativistic arguments, thus emphasized once more the necessity of distinguishing, in the study of atomic phenomena, between the proper measuring instruments which serve to define the reference frame and those parts which are to be regarded as objects under investigation and in the account of which quantum effects cannot be disregarded. Notwithstanding the most suggestive confirmation of the soundness and wide scope of the quantum-mechanical way of description, Einstein nevertheless, in a following conversation with me, expressed a feeling of disquietude as regards the apparent lack of firmly laid down principles for the explanation of nature, in which all could agree. From my viewpoint, however, I could only answer that, in dealing with the task of bringing order into an entirely new field of experience, we could hardly trust in any accustomed principles, however broad, apart from the demand of avoiding logical inconsistencies and, in this respect, the mathematical formalism of quantum mechanics should surely meet all requirements.

The Solvay meeting in 1930 was the last occasion where, in common discussions with Einstein, we could benefit from the stimulating and mediating influence of Ehrenfest, but shortly before his deeply deplored death in 1933 he told me that Einstein was far from satisfied and with his usual acuteness had discerned new aspects of the situation which strengthened his critical attitude. In fact, by further examining the possibilities for the application of a balance arrangement, Einstein had perceived alternative procedures which, even if they did not allow the use he originally intended, might seem to enhance the paradoxes beyond the possibilities of logical solution. Thus, Einstein had pointed out that, after a preliminary weighing of the box with the clock and the subsequent escape of the photon, one was still left with the choice of either repeating the weighing or opening the box and comparing the reading of the clock with the standard time scale. Consequently, we are at this stage still free to choose whether we want to draw conclusions either about the energy of the photon or about the moment when it left the box. Without in any way interfering with the photon between its escape and its later interaction with other suitable measuring instruments, we are, thus, able to make accurate predictions pertaining *either* to the moment of its arrival *or* to the amount of energy liberated by its absorption. Since, however, according to the quantum-mechanical formalism, the specification of the state of an isolated particle cannot involve both a well-defined connection with the time scale and an accurate fixation of the energy, it might thus appear as if this formalism did not offer the means of an adequate description.

Once more Einstein's searching spirit had elicited a peculiar aspect of the situation in quantum theory, which in a most striking manner illustrated how far we have here transcended customary explanation of natural phenomena. Still, I could not agree with the trend of his remarks as reported by Ehrenfest. In my opinion, there could be no other way to deem a logically consistent mathematical formalism as inadequate than by demonstrating the departure of its consequences from experience or by proving that its predictions did not exhaust the possibilities of observation, and Einstein's argumentation could be directed to neither of these ends. In fact, we must realize that in the problem in question we are not dealing with a *single* specified experimental arrangement, but are referring to *two* different, mutually exclusive arrangements. In the one, the balance together with another piece of apparatus like a spectrometer is used for the study of the energy transfer by a photon; in the other, a shutter regulated by a standardized clock together with another apparatus of similar kind, accurately timed relatively to the clock, is used for the study of the time of propagation of a photon over a given distance. In both these cases, as also assumed by Einstein, the observable effects are expected to be in complete conformity with the predictions of the theory.

The problem again emphasizes the necessity of considering the *whole* experimental arrangement, the specification of which is imperative for any well-defined application of the quantum-mechanical formalism. Incidentally, it may be added that paradoxes of the kind contemplated by Einstein are encountered also in such simple arrangements as sketched in Figure 5. In fact, after a preliminary measurement of the momentum of the diaphragm, we are in principle offered the choice, when an electron or photon has passed through the slit, either to repeat the momentum measurement or to control the position of the diaphragm and, thus, to make predictions pertaining to alternative subsequent observations. It may also be added that it obviously can make no difference, as regards observable effects obtainable by a definite experimental arrangement, whether our plans of constructing or handling the instruments are fixed beforehand or whether we prefer to postpone the completion of our planning until a later moment when the particle is already on its way from one instrument to another.

In the quantum-mechanical description our freedom of constructing and handling the experimental arrangement finds its proper expression in the possibility of choosing the classically defined parameters entering in any proper application of the formalism. Indeed, in all such respects quantum mechanics exhibits a correspondence with the state of affairs familiar from classical physics, which is as close as possible when considering the individuality inherent in the quantum phenomena. Just in helping to bring out this point so clearly, Einstein's concern had therefore again been a most welcome incitement to explore the essential aspects of the situation.

The next Solvay meeting in 1933 was devoted to the problems of the structure and properties of atomic nuclei, in which field such great advances were made just in that period owing to the experimental discoveries as well as to new fruitful applications of quantum mechanics. It need in this connection hardly be recalled that just the evidence obtained by the study of artificial nuclear transformations gave a most direct test of Einstein's fundamental law regarding the equivalence of mass and energy, which was to prove an evermore important guide for researches in nuclear physics. It may also be mentioned how Einstein's intuitive recognition of the intimate relationship between the law of radioactive transformations and the probability rules governing individual radiation effects (cf. p. 35) was confirmed by the quantum-mechanical explanation of spontaneous nuclear disintegrations. In fact, we are here dealing with a typical example of the statistical mode of description, and the complementary relationship between energy-momentum conservation and time-space coordination is most strikingly exhibited in the well-known paradox of particle penetration through potential barriers.

Einstein himself did not attend this meeting, which took place at a time darkened by the tragic developments in the political world which were to influence his fate so deeply and add so greatly to his burdens in the service of humanity. A few months earlier, on a visit to Princeton where Einstein was then guest of the newly founded Institute for Advanced Study to which he soon after became permanently attached, I had, however, opportunity to talk with him again about the epistemological aspects of atomic physics, but the difference between our ways of approach and expression still presented obstacles to mutual understanding. While, so far, relatively few persons had taken part in the discussions reported in this article, Einstein's critical attitude towards the views on quantum theory adhered to by many physicists was soon after brought to public attention through a paper [11] with the title "Can Quantum-Mechanical Description of Physical Reality Be Considered Complete?," published in 1935 by Einstein, Podolsky and Rosen.

[11] A. Einstein, B. Podolsky and N. Rosen, *Phys. Rev.*, 47, 777 (1935).

The argumentation in this paper is based on a criterion which the authors express in the following sentence: "If, without in any way disturbing a system, we can predict with certainty (i.e., with probability equal to unity) the value of a physical quantity, then there exists an element of physical reality corresponding to this physical quantity." By an elegant exposition of the consequences of the quantum-mechanical formalism as regards the representation of a state of a system, consisting of two parts which have been in interaction for a limited time interval, it is next shown that different quantities, the fixation of which cannot be combined in the representation of one of the partial systems, can nevertheless be predicted by measurements pertaining to the other partial system. According to their criterion, the authors therefore conclude that quantum mechanics does not "provide a complete description of the physical reality," and they express their belief that it should be possible to develop a more adequate account of the phenomena.

Due to the lucidity and apparently incontestable character of the argument, the paper of Einstein, Podolsky and Rosen created a stir among physicists and has played a large role in general philosophical discussion. Certainly the issue is of a very subtle character and suited to emphasize how far, in quantum theory, we are beyond the reach of pictorial visualization. It will be seen, however, that we are here dealing with problems of just the same kind as those raised by Einstein in previous discussions, and, in an article which appeared a few months later,[12] I tried to show that from the point of view of complementarity the apparent inconsistencies were completely removed. The trend of the argumentation was in substance the same as that exposed in the foregoing pages, but the aim of recalling the way in which the situation was discussed at that time may be an apology for citing certain passages from my article.

Thus, after referring to the conclusions derived by Einstein, Podolsky and Rosen on the basis of their criterion, I wrote:

Such an argumentation, however, would hardly seem suited to affect the soundness of quantum-mechanical description, which is based on a coherent mathematical formalism covering automatically any procedure of measurement like that indicated. The apparent contradiction in fact discloses only an essential inadequacy of the customary viewpoint of natural philosophy for a rational account of physical phenomena of the type with which we are concerned in quantum mechanics. Indeed the *finite interaction between object and measuring agencies* conditioned by the very existence of the quantum of action entails—because of the impossibility of controlling the reaction of the object on the measuring

12 N. Bohr, *Phys. Rev.*, *48*, 696 (1935).

instruments, if these are to serve their purpose—the necessity of a final renunciation of the classical ideal of causality and a radical revision of our
attitude towards the problem of physical reality. In fact, as we shall see,
a criterion of reality like that proposed by the named authors contains—
however cautious its formulation may appear—an essential ambiguity when
it is applied to the actual problems with which we are here concerned.

As regards the special problem treated by Einstein, Podolsky and
Rosen, it was next shown that the consequences of the formalism
as regards the representation of the state of a system consisting of two
interacting atomic objects correspond to the simple arguments mentioned in the preceding in connection with the discussion of the experimental arrangements suited for the study of complementary phenomena. In fact, although any pair q and p of conjugate space and
momentum variables obeys the rule of non-commutative multiplica-
tion expressed by (2), and can thus only be fixed with reciprocal latitudes given by (3), the difference $q_1 - q_2$ between two space-co-
ordinates referring to the constituents of the system will commute
with the sum $p_1 + p_2$ of the corresponding momentum components,
as follows directly from the commutability of q_1 with p_2 and q_2
with p_1. Both $q_1 - q_2$ and $p_1 + p_2$ can, therefore, be accurately
fixed in a state of the complex system and, consequently, we can
predict the values of either q_1 or p_1 if either q_2 or p_2, respectively, is
determined by direct measurements. If, for the two parts of the system, we take a particle and a diaphragm, like that sketched in Figure
5, we see that the possibilities of specifying the state of the particle
by measurements on the diaphragm just correspond to the situation
described on p. 48 and further discussed on p. 57, where it was mentioned that, after the particle has passed through the diaphragm, we
have in principle the choice of measuring either the position of the
diaphragm or its momentum and, in each case, making predictions as
to subsequent observations pertaining to the particle. As repeatedly
stressed, the principal point here is that such measurements demand
mutually exclusive experimental arrangements.

The argumentation of the article was summarized in the following
passage:

From our point of view we now see that the wording of the above-
mentioned criterion of physical reality proposed by Einstein, Podolsky
and Rosen contains an ambiguity as regards the meaning of the expression "without in any way disturbing a system." Of course there is in a
case like that just considered no question of a mechanical disturbance of
the system under investigation during the last critical stage of the measuring procedure. But even at this stage there is essentially the question of

an influence on the very conditions which define the possible types of predictions regarding the future behaviour of the system. Since these conditions constitute an inherent element of the description of any phenomenon to which the term "physical reality" can be properly attached, we see that the argumentation of the mentioned authors does not justify their conclusion that quantum-mechanical description is essentially incomplete. On the contrary, this description, as appears from the preceding discussion, may be characterized as a rational utilization of all possibilities of unambiguous interpretation of measurements, compatible with the finite and uncontrollable interaction between the objects and the measuring instruments in the field of quantum theory. In fact, it is only the mutual exclusion of any two experimental procedures, permitting the unambiguous definition of complementary physical quantities, which provides room for new physical laws, the coexistence of which might at first sight appear irreconcilable with the basic principles of science. It is just this entirely new situation as regards the description of physical phenomena that the notion of *complementarity* aims at characterizing.

Rereading these passages, I am deeply aware of the inefficiency of expression which must have made it very difficult to appreciate the trend of the argumentation aiming to bring out the essential ambiguity involved in a reference to physical attributes of objects when dealing with phenomena where no sharp distinction can be made between the behaviour of the objects themselves and their interaction with the measuring instruments. I hope, however, that the present account of the discussions with Einstein in the foregoing years, which contributed so greatly to make us familiar with the situation in quantum physics, may give a clearer impression of the necessity of a radical revision of basic principles for physical explanation in order to restore logical order in this field of experience.

Einstein's own views at that time are presented in an article "Physics and Reality," published in 1936 in the *Journal of the Franklin Institute.*[13] Starting from a most illuminating exposition of the gradual development of the fundamental principles in the theories of classical physics and their relation to the problem of physical reality, Einstein here argues that the quantum-mechanical description is to be considered merely as a means of accounting for the average behaviour of a large number of atomic systems, and his attitude to the belief that it should offer an exhaustive description of the individual phenomena is expressed in the following words: "To believe this is logically possible without contradiction; but it is so very contrary to my scientific instinct that I cannot forego the search for a more complete conception."

[13] A. Einstein, *J. Franklin Inst.*, 221, 349 (1936).

Even if such an attitude might seem well balanced in itself, it nevertheless implies a rejection of the whole argumentation exposed in the preceding, aiming to show that, in quantum mechanics, we are not dealing with an arbitrary renunciation of a more detailed analysis of atomic phenomena, but with a recognition that such an analysis is *in principle* excluded. The peculiar individuality of the quantum effects presents us, as regards the comprehension of well-defined evidence, with a novel situation unforeseen in classical physics and irreconcilable with conventional ideas suited for our orientation and adjustment to ordinary experience. It is in this respect that quantum theory has called for a renewed revision of the foundation for the unambiguous use of elementary concepts as a further step in the development which, since the advent of relativity theory, has been so characteristic of modern science.

In the following years, the more philosophical aspects of the situation in atomic physics aroused the interest of ever larger circles and were, in particular, discussed at the Second International Congress for the Unity of Science in Copenhagen in July 1936. In a lecture on this occasion,[14] I tried especially to stress the analogy in epistemological respects between the limitation imposed on the causal description in atomic physics and situations met with in other fields of knowledge. A principal purpose of such parallels was to call attention to the necessity in many domains of general human interest of facing problems of a similar kind as those which had arisen in quantum theory and thereby to give a more familiar background for the apparently extravagant way of expression which physicists have developed to cope with their acute difficulties.

Besides the complementary features conspicuous in psychology and already touched upon (cf. p. 52), examples of such relationships can also be traced in biology, especially as regards the comparison between mechanistic and vitalistic viewpoints. Just with respect to the observational problem, this last question had previously been the subject of an address to the International Congress on Light Therapy held in Copenhagen in 1932,[15] where it was incidentally pointed out that even the psycho-physical parallelism as envisaged by Leibniz and Spinoza has obtained a wider scope through the development of atomic physics, which forces us to an attitude towards the problem

---

[14] N. Bohr, *Philosophy of Science*, **4**, 289 (1937).

[15] IIe Congrès international de la Lumière, Copenhague 1932 (reprinted in this collection, p. 3).

of explanation recalling ancient wisdom, that when searching for harmony in life one must never forget that in the drama of existence we are ourselves both actors and spectators. Utterances of this kind would naturally in many minds evoke the impression of an underlying mysticism foreign to the spirit of science; at the above-mentioned Congress in 1936 I therefore tried to clear up such misunderstandings and to explain that the only question was an endeavour to clarify the conditions, in each field of knowledge, for the analysis and synthesis of experience.[14] Yet, I am afraid that I had in this respect only little success in convincing my listeners, for whom the dissent among the physicists themselves was naturally a cáuse of scepticism about the necessity of going so far in renouncing customary demands as regards the explanation of natural phenomena. Not least through a new discussion with Einstein in Princeton in 1937, where we did not get beyond a humourous contest concerning which side Spinoza would have taken if he had lived to see the development of our days, I was strongly reminded of the importance of utmost caution in all questions of terminology and dialectics.

These aspects of the situation were especially discussed at a meeting in Warsaw in 1938, arranged by the International Institute of Intellectual Co-operation of the League of Nations.[16] The preceding years had seen great progress in quantum physics owing to a number of fundamental discoveries regarding the constitution and properties of atomic nuclei as well as important developments of the mathematical formalism taking the requirements of relativity theory into account. In the last respect, Dirac's ingenious quantum theory of the electron offered a most striking illustration of the power and fertility of the general quantum-mechanical way of description. In the phenomena of creation and annihilation of electron pairs we have in fact to do with new fundamental features of atomicity, which are intimately connected with the non-classical aspects of quantum statistics expressed in the exclusion principle, and which have demanded a still more far-reaching renunciation of explanation in terms of a pictorial representation.

Meanwhile, the discussion of the epistemological problems in atomic physics attracted as much attention as ever and, in commenting on Einstein's views as regards the incompleteness of the quantum-mechanical mode of description, I entered more directly on questions of terminology. In this connection I warned especially against phrases, often found in the physical literature, such as "dis-

[16] *New Theories in Physics* (Paris 1938), 11.

turbing of phenomena by observation" or "creating physical attributes to atomic objects by measurements." Such phrases, which may serve to remind of the apparent paradoxes in quantum theory, are at the same time apt to cause confusion, since words like "phenomena" and "observations," just as "attributes" and "measurements," are used in a way hardly compatible with common language and practical definition.

As a more appropriate way of expression I advocated the application of the word *phenomenon* exclusively to refer to the observations obtained under specified circumstances, including an account of the whole experimental arrangement. In such terminology, the observational problem is free of any special intricacy since, in actual experiments, all observations are expressed by unambiguous statements referring, for instance, to the registration of the point at which an electron arrives at a photographic plate. Moreover, speaking in such a way is just suited to emphasize that the appropriate physical interpretation of the symbolic quantum-mechanical formalism amounts only to predictions, of determinate or statistical character, pertaining to individual phenomena appearing under conditions defined by classical physical concepts.

Notwithstanding all differences between the physical problems which have given rise to the development of relativity theory and quantum theory, respectively, a comparison of purely logical aspects of relativistic and complementary argumentation reveals striking similarities as regards the renunciation of the absolute significance of conventional physical attributes of objects. Also, the neglect of the atomic constitution of the measuring instruments themselves, in the account of actual experience, is equally characteristic of the applications of relativity and quantum theory. Thus, the smallness of the quantum of action compared with the actions involved in usual experience, including the arranging and handling of physical apparatus, is as essential in atomic physics as is the enormous number of atoms composing the world in the general theory of relativity which, as is often pointed out, demands that dimensions of apparatus for measuring angles can be made small compared with the radius of curvature of space.

In the Warsaw lecture, I commented upon the use of not directly visualizable symbolism in relativity and quantum theory in the following way:

Even the formalisms, which in both theories within their scope offer adequate means of comprehending all conceivable experience, exhibit deepgoing analogies. In fact, the astounding simplicity of the generalization of classical physical theories, which are obtained by the use of multidimensional geometry and non-commutative algebra, respectively, rests in both cases essentially on the introduction of the conventional symbol $\sqrt{-1}$. The abstract character of the formalisms concerned is indeed, on closer examination, as typical of relativity theory as it is of quantum mechanics, and it is in this respect purely a matter of tradition if the former theory is considered as a completion of classical physics rather than as a first fundamental step in the thoroughgoing revision of our conceptual means of comparing observations, which the modern development of physics has forced upon us.

It is, of course, true that in atomic physics we are confronted with a number of unsolved fundamental problems, especially as regards the intimate relationship between the elementary unit of electric charge and the universal quantum of action; but these problems are no more connected with the epistemological points here discussed than is the adequacy of relativistic argumentation with the issue of thus far unsolved problems of cosmology. Both in relativity and in quantum theory we are concerned with new aspects of scientific analysis and synthesis and, in this connection, it is interesting to note that, even in the great epoch of critical philosophy in the former century, there was only question to what extent *a priori* arguments could be given for the adequacy of space-time coordination and causal connection of experience, but never question of rational generalizations or inherent limitations of such categories of human thinking.

Although in more recent years I have had several occasions of meeting Einstein, the continued discussions, from which I always have received new impulses, have so far not led to a common view about the epistemological problems in atomic physics, and our opposing views are perhaps most clearly stated in a recent issue of *Dialectica*,[17] bringing a general discussion of these problems. Realizing, however, the many obstacles for mutual understanding as regards a matter where approach and background must influence everyone's attitude, I have welcomed this opportunity of a broader exposition of the development by which, to my mind, a veritable crisis in physical science has been overcome. The lesson we have hereby received would seem to have brought us a decisive step further in the neverending struggle for harmony between content and form, and taught us once again that no content can be grasped without a formal frame and that any form, however useful it has hitherto proved, may be found to be too narrow to comprehend new experience.

Surely, in a situation like this, where it has been difficult to reach mutual understanding not only between philosophers and physicists

[17] N. Bohr, *Dialectica*, *1*, 312 (1948).

but even between physicists of different schools, the difficulties have their root not seldom in the preference for a certain use of language suggesting itself from the different lines of approach. In the Institute in Copenhagen, where through those years a number of young physicists from various countries came together for discussions, we used, when in trouble, often to comfort ourselves with jokes, among them the old saying of the two kinds of truth. To the one kind belong statements so simple and clear that the opposite assertion obviously could not be defended. The other kind, the so-called "deep truths," are statements in which the opposite also contains deep truth. Now, the development in a new field will usually pass through stages in which chaos becomes gradually replaced by order; but it is not least in the intermediate stage where deep truth prevails that the work is really exciting and inspires the imagination to search for a firmer hold. For such endeavours of seeking the proper balance between seriousness and humour, Einstein's own personality stands as a great example and, when expressing my belief that through a singularly fruitful cooperation of a whole generation of physicists we are nearing the goal where logical order to a large extent allows us to avoid deep truth, I hope that it will be taken in his spirit and may serve as an apology for several utterances in the preceding pages.

The discussions with Einstein which have formed the theme of this article have extended over many years which have witnessed great progress in the field of atomic physics. Whether our actual meetings have been of short or long duration, they have always left a deep and lasting impression on my mind, and when writing this report I have, so-to-speak, been arguing with Einstein all the time, even in discussing topics apparently far removed from the special problems under debate at our meetings. As regards the account of the conversations I am, of course, aware that I am relying only on my own memory, just as I am prepared for the possibility that many features of the development of quantum theory, in which Einstein has played so large a part, may appear to himself in a different light. I trust, however, that I have not failed in conveying a proper impression of how much it has meant to me to be able to benefit from the inspiration which we all derive from every contact with Einstein.

Unity
of
Knowledge

1954

Before trying to answer the question to what extent we may speak of unity of knowledge, we may ask for the meaning of the word knowledge itself. It is not my intention to enter into an academic philosophical discourse for which I would hardly possess the required scholarship. Every scientist, however, is constantly confronted with the problem of objective description of experience, by which we mean unambiguous communication. Our basic tool is, of course, plain language which serves the needs of practical life and social intercourse. We shall not be concerned here with the origins of such language, but with its scope in scientific communication, and especially with the problem of how objectivity may be retained during the growth of experience beyond the events of daily life.

The main point to realize is that all knowledge presents itself within a conceptual framework adapted to account for previous experience and that any such frame may prove too narrow to comprehend new experiences. Scientific research in many domains of knowledge has indeed time and again proved the necessity of abandoning or remoulding points of view which, because of their fruitfulness and apparently unrestricted applicability, were regarded as indispensable for rational explanation. Although such developments have been initiated by special studies, they entail a general lesson of importance for the problem of unity of knowledge. In fact, the widening of the conceptual framework not only has served to restore order within the respective branches of knowledge, but has also disclosed analogies in our position with respect to analysis and synthesis of experience in apparently separated domains of knowledge, suggesting the possibility of an ever more embracing objective description.

When speaking of a conceptual framework, we refer merely to the unambiguous logical representation of relations between experiences. This attitude is also apparent in the historical development in which formal logic is no longer sharply distinguished from studies of semantics or even philological syntax. A special role is played by mathematics which has contributed so decisively to the development of logical thinking, and which by its well-defined abstractions offers invaluable help in expressing harmonious relationships. Still, in our discussion, we shall not consider pure mathematics as a separate branch of knowledge, but rather as a refinement of general language, supplementing it with appropriate tools to represent relations for which ordinary verbal expression is imprecise or cumbersome. In this connection, it may be stressed that, just by avoiding the reference to the conscious subject which infiltrates daily language, the use of mathematical symbols secures the unambiguity of definition required for objective description.

The development of the so-called exact sciences, characterized by the establishing of numerical relationships between measurements, has indeed been decisively furthered by abstract mathematical methods originating from detached pursuit of generalizing logical constructions. This situation is especially illustrated in physics which was originally understood as all knowledge concerning that nature of which we ourselves are part, but gradually came to mean the study of the elementary laws governing the properties of inanimate matter. The necessity, even within this comparatively simple theme, of paying constant attention to the problem of objective description has deeply influenced the attitude of philosophical schools through the ages. In our day, the exploration of new fields of experience has disclosed unsuspected presuppositions for the unambiguous application of some of our most elementary concepts and thereby given us an epistemological lesson with bearings on problems far beyond the domain of physical science. It may therefore be convenient to start our discussion with a brief account of this development.

It would carry us too far to recall in detail how, with the elimination of mythical cosmological ideas and arguments referring to the purpose for our own actions, a consistent scheme of mechanics was built up on the basis of Galileo's pioneering work and reached such completion through Newton's mastery. Above all, the principles of Newtonian mechanics meant a far-reaching clarification of the problem of cause and effect by permitting, from the state of a physical system defined at a given instant by measurable quantities, the prediction of its state at any subsequent time. It is well known how a deterministic or causal account of this kind led to the mechanical conception of nature and came to stand as an ideal of scientific explanation in all domains of knowledge, irrespective of the way knowledge is obtained. In this connection, therefore, it is important that the study of wider fields of physical experience has revealed the necessity of a closer consideration of the observational problem.

Within its large field of application, classical mechanics presents an objective description in the sense that it is based on a well-defined use of pictures and ideas referring to the events of daily life. Still, however rational the idealizations used in Newtonian mechanics might appear, they actually went far beyond the range of experience to which our elementary concepts are adapted. Thus, the adequate use of the very notions of absolute space and time is inherently connected with the practically instantaneous propagation of light, which allows us to locate the bodies around us independently of their velocities and to arrange events in a unique time sequence. However, the attempt to develop a consistent account of electromagnetic and optical phenomena revealed that observers moving relative to each other with large velocities will coordinate events differently. Not only may such observers take a different view of shapes and positions of rigid bodies, but events at separate points of space which to one observer appear as simultaneous may be judged by another as occurring at different times.

Far from giving rise to confusion and complication, the exploration of the extent to which the account of physical phenomena depends on the standpoint of the observer proved an invaluable guide in tracing general physical laws common to all observers. Retaining the idea of determinism, but relying only on relations between unambiguous measurements referring ultimately to coincidences of events, Einstein succeeded in remoulding and generalizing the whole edifice of classical physics and in lending to our world picture a unity surpassing all previous expectations. In the general theory of relativity, the description is based on a curved four-dimensional space-time metric which automatically accounts for gravitational effects and the singular role of the speed of light signals representing an upper limit for any consistent use of the physical concept of velocity. The introduction of such unfamiliar but well-defined mathematical abstractions in no way implies ambiguity but rather offers an instructive illustration of how a widening of the conceptual framework affords the appropriate means of eliminating subjective elements and enlarging the scope of objective description.

New, unsuspected aspects of the observational problem were disclosed by the exploration of the atomic constitution of matter. As is well known, the idea of a limited divisibility of substances, introduced to explain the persistence of their characteristic properties in spite of the variety of natural phenomena, goes back to antiquity. Still, almost to our day, such views were regarded as essentially hypothetical in the sense that they seemed inaccessible to direct confirmation by observation because of the coarseness of our sense organs and tools, themselves composed of innumerable atoms. Nevertheless, with the great progress in chemistry and physics in the last centuries, atomic ideas proved increasingly fruitful. In particular, the direct application of classical mechanics to the interaction of atoms and molecules during their incessant motions led to a general understanding of the principles of thermodynamics.

In this century, the study of newly discovered properties of matter such as natural radioactivity has convincingly confirmed the foundations of atomic theory. In particular, through the development of amplification devices, it has been possible to study phenomena essentially dependent on single atoms, and even to obtain extensive knowledge of the structure of atomic systems. The first step was the recognition of the electron as a common constituent of all substances, and an essential completion of our ideas of atomic constitution was obtained by Rutherford's discovery of the atomic nucleus which contains within an extremely small volume almost the whole mass of the atom. The invariability of the properties of the elements in ordinary physical and chemical processes is directly explained by the circumstance that in such processes, although the electron binding may be largely influenced, the nucleus remains unaltered. With his demonstration of the transmutability of atomic nuclei by more powerful agencies, Rutherford, however, opened a quite new field of research, often referred to as modern alchemy, which, as is well known, was eventually to lead to the possibility of releasing immense amounts of energy stored in atomic nuclei.

Although many fundamental properties of matter were explained by the simple picture of the atom, it was evident from the beginning that classical ideas of mechanics and electromagnetism did not suffice to account for the essential stability of atomic structures, as exhibited by the specific properties of the elements. However, a clue to the elucidation of this problem was afforded by the discovery of the universal quantum of action to which Planck was led in the first year of our century by his penetrating analysis of the laws of thermal radiation. This discovery revealed in atomic processes a feature of wholeness quite foreign to the mechanical conception of nature, and made it evident that the classical physical theories are idealizations valid only in the description of phenomena in the analysis of which all actions are sufficiently large to permit the neglect of the quantum. While this condition is amply fulfilled in phenomena on the ordinary scale, we meet in atomic phenomena regularities of quite a new kind, defying deterministic pictorial description.

A rational generalization of classical physics, allowing for the existence of the quantum but retaining the unambiguous interpretation of the experimental evidence defining the inertial mass and electric charge of the electron and the nucleus, presented a very difficult task. By concerted efforts of a whole generation of theoretical physicists, a consistent and, within a wide scope, exhaustive description of atomic phenomena was, however, gradually developed. This description makes use of a mathematical formalism in which the variables in the classical physical theories are replaced by symbols subject to a noncommutable algorism involving Planck's constant. Owing to the very character of such mathematical abstractions, the formalism does not allow pictorial interpretation on accustomed lines, but aims directly at establishing relations between observations obtained under well-defined conditions. Corresponding to the circumstance that different individual quantum processes may take place in a given experimental arrangement, these relations are of an inherently statistic character.

By means of the quantum mechanical formalism, a detailed account of an immense amount of experimental evidence regarding the physical and chemical properties of matter has been achieved. Moreover, by adapting the formalism to the exigencies of relativistic invariance, it has been possible, within wide limits, to order the rapidly growing new knowledge concerning the properties of elementary particles and the constitution of atomic nuclei. Notwithstanding the astounding power of quantum mechanics, the radical departure from accustomed physical explanation, and especially the renunciation of the very idea of determinism, has given rise to doubts in the minds of many physicists and philosophers as to whether we are here dealing with a temporary expedient or are confronted with an irrevocable step as regards objective description. The clarification of this problem has actually demanded a radical revision of the fundamentals to the description and comprehension of physical experience.

In this context, we must recognize above all that, even when the phenomena transcend the scope of classical physical theories, the account of the experimental arrangement and the recording of observations must be given in plain language, suitably supplemented by technical physical terminology. This is a clear logical demand, since the very word "experiment" refers to a situation where we can tell others what we have done and what we have learned. However, the fundamental difference with respect to the analysis of phenomena in classical and in quantum physics is that in the former the interaction between the objects and the measuring instruments may be neglected or compensated for, while in the latter this interaction forms an integral part of the phenomena. The essential wholeness of a proper quantum phenomenon finds indeed logical expression in the circumstance that any attempt at its well-defined subdivision would require a change in the experimental arrangement incompatible with the appearance of the phenomenon itself.

In particular, the impossibility of a separate control of the interaction between the atomic objects and the instruments indispensable for the definition of the experimental conditions prevents the unrestricted combination of space-time coordination and dynamical conservation laws on which the deterministic description in classical physics rests. In fact, any unambiguous use of the concepts of space and time refers to an experimental arrangement involving a transfer of momentum and energy, uncontrollable in principle, to fixed scales and synchronized clocks which are required for the definition of the reference frame. Conversely, the account of phenomena which are characterized by the laws of conservation of momentum and energy involves in principle a renunciation of detailed space-time coordination. These circumstances find quantitative expression in Heisenberg's indeterminacy relations which specify the reciprocal latitude for the fixation of kinematical and dynamical variables in the definition of the state of a physical system. In accordance with the character of the quantum mechanical formalism, such relations cannot, however, be interpreted in terms of attributes of objects referring to classical pictures, but we are here dealing with the mutually exclusive conditions for the unambiguous use of the very concepts of space and time on the one hand, and of dynamical conservation laws on the other.

In this context, one sometimes speaks of "disturbance of phenomena by observation" or "creation of physical attributes to atomic objects by measurements." Such phrases, however, are apt to cause confusion, since words like phenomena and observation, just as attributes and measurements, are here used in a way incompatible with common language and practical definition. On the lines of objective description, it is indeed more appropriate to use the word phenomenon to refer only to observations obtained under circumstances whose description includes an account of the whole experimental arrangement. In such terminology, the observational problem in quantum physics is deprived of any special intricacy and we are, moreover, directly reminded that every atomic phenomenon is closed in the sense that its observation is based on registrations obtained by means of suitable amplification devices with irreversible functioning such as, for example, permanent marks on a photographic plate, caused by the penetration of electrons into the emulsion. In this connection, it is important to realize that the quantum-mechanical formalism permits well-defined applications referring only to such closed phenomena. Also in this respect it represents a rational generalization of classical physics in which every stage of the course of events is described by measurable quantities.

The freedom of experimentation, presupposed in classical physics, is of course retained and corresponds to the free choice of experimental arrangements for which the mathematical structure of the quantum mechanical formalism offers the appropriate latitude. The circumstance that, in general, one and the same experimental arrangement may yield different recordings is sometimes picturesquely described as a "choice of nature" between such possibilities. Needless to say, such a phrase implies no allusion to a personification of nature, but simply points to the impossibility of ascertaining on accustomed lines directives for the course of a closed indivisible phenomenon. Here, logical approach cannot go beyond the deduction of the relative probabilities for the appearance of the individual phenomena under given experimental conditions. In this respect, quantum mechanics presents a consistent generalization of deterministic mechanical description which it embraces as an asymptotic limit in the case of physical phenomena on a scale sufficiently large to allow the neglect of the quantum of action.

A most conspicuous characteristic of atomic physics is the novel relationship between phenomena observed under experimental conditions demanding different elementary concepts for their description. Indeed, however contrasting such experiences might appear when attempting to picture a course of atomic processes on classical lines, they have to be considered as complementary in the sense that they represent equally essential knowledge about atomic systems and together exhaust this knowledge. The notion of complementarity does in no way involve a departure from our position as detached observers of nature, but must be regarded as the logical expression of our situation as regards objective description in this field of experience. The recognition that the interaction between the measuring tools and the physical systems under investigation constitutes an integral part of quantum phenomena has not only revealed an unsuspected limitation of the mechanical conception of nature, as characterized by attribution of separate properties to physical systems, but has forced us, in the ordering of experience, to pay proper attention to the conditions of observation.

Returning to the much debated question of what has to be demanded of a physical explanation, one must keep in mind that classical mechanics had already implied the renunciation of a cause for uniform motion and furthermore that relativity theory has taught us how arguments of invariance and equivalence must be treated as categories of rational explanation. Similarly, in the complementary description of quantum physics, we have to do with a further self-consistent generalization which permits the inclusion of regularities decisive for the account of fundamental properties of matter, but which transcends the scope of deterministic description. The history of physical science thus demonstrates how the exploration of ever wider fields of experience, in revealing unsuspected limitations of accustomed ideas, indicates new ways of restoring logical order. As we shall now proceed to show, the epistemological lesson contained in the development of atomic physics reminds us of similar situations with respect to the description and comprehension of experience far beyond the borders of physical science, and allows us to trace common features promoting the search for unity of knowledge.

The first problem with which we are confronted when leaving the proper domain of physics is the question of the place of living organisms in the description of natural phenomena. Originally, no sharp distinction between animate and inanimate matter was made, and it is well known that Aristotle, in stressing the wholeness of the individual organisms, opposed the views of the atomists, and even in the discussion of the foundations of mechanics retained ideas like purpose and potency. However, as a result of the great discoveries in anatomy and physiology at the time of the Renaissance, and especially of the advent of classical mechanics in the deterministic description of which any reference to purpose is eliminated, a completely mechanistic conception of nature suggested itself, and a large number of organic functions could in fact be accounted for by the same physical and chemical properties of matter which found far-reaching explanation on simple atomic ideas. It is true that the structure and functioning of organisms involve an ordering of atomic processes which has sometimes seemed difficult to reconcile with the laws of thermodynamics, implying a steady approach towards disorder among the atoms constituting an isolated physical system. If, however, sufficient account is taken of the circumstance that the free energy necessary to maintain and develop organic systems is continually supplied from their surroundings by nutrition and respiration, it becomes clear that there is in such respect no question of any violation of general physical laws.

In the last decades, great advances have been achieved in our knowledge of the structure and functioning of organisms, and in particular it has become evident that quantum regularities in many respects here play a fundamental role. Not only are such regularities basic of the remarkable stability of the highly complex molecular structures which form the essential constituents of the cells responsible for the hereditary properties of the species, but research on mutations produced by exposing organisms to penetrating radiation offers a striking application of the statistical laws of quantum physics. Also, the sensitivity of perceptive organs, so important for the integrity of the organisms, has been found to approach the level of individual quantum processes, and amplification mechanisms play an important part especially in the transmission of nervous messages. The whole development has again, although in a novel manner, brought the mechanistic approach to biological problems to the foreground, but at the same time the question has become acute as to whether a comparison between the organisms and highly complex and refined systems, such as modern industrial constructions or electronic calculation machines, offers the proper basis for an objective description of the self-regulating entities which living organisms present.

Returning to the general epistemological lesson which atomic physics has given us, we must in the first place realize that the closed processes studied in quantum physics are not directly analogous to biological functions for the maintenance of which a continual exchange of matter and energy between the organism and the environments is required. Moreover, any experimental arrangement which would permit control of such functions to the extent demanded for their well-defined description in physical terms would be prohibitive to the free display of life. This very circumstance, however, suggests an attitude to the problem of organic life providing a more appropriate balance between a mechanistic and a finalistic approach. In fact, just as the quantum of action appears in the account of atomic phenomena as an element for which an explanation is neither possible nor required, the notion of life is elementary in biological science where, in the existence and evolution of living organisms, we are concerned with manifestations of possibilities in that nature to which we belong rather than with the outcome of experiments which we can ourselves perform. Actually, we must recognize that the requirements of objective description, in tendency at least, are fulfilled by the characteristic complementary way in which arguments based on the full resources of physical and chemical science, and concepts directly referring to the integrity of the organism transcending the scope of these sciences, are practically used in biological research. The main point is that only by renouncing an explanation of life in the ordinary sense do we gain a possibility of taking into account its characteristics.

Of course, in biology just as in physics, we retain our position as detached observers, and the question is only that of the different conditions for the logical comprehension of experience. This applies also to the study of the innate and conditioned behaviour of animals and man to which psychological concepts readily lend themselves. Even in an allegedly behaviouristic approach, it is hardly possible to avoid such concepts, and the very idea of consciousness presents itself when we deal with behaviour of so high a degree of complexity that its description virtually involves introspection on the part of the individual organism. We have here to do with mutually exclusive applications of the words instinct and reason, illustrated by the degree to which instinctive behaviour is suppressed in human societies. Although we meet in trying to account for the state of our mind ever greater difficulties as regards observational detachment, it is still possible to uphold the requirements of objective description to a great extent even in human psychology. In this connection, it is interesting to note that, while in the early stages of physical science one could directly rely on such features of the events of daily life which permitted a simple causal account, an essentially complementary description of the content of our mind has been used since the origin of languages. In fact, the rich terminology adapted to such communication does not point to an unbroken course of events, but rather to mutually exclusive experiences characterized by different separations between the content on which attention is focused and the background indicated by the word ourselves.

An especially striking example is offered by the relationship between situations in which we ponder on the motives for our actions and in which we experience a feeling of volition. In normal life, such shifting of the separation is more or less intuitively recognized, but symptoms characterized as "confusion of the egos," which may lead to dissolution of the personality, are well known in psychiatry. The use of apparently contrasting attributes referring to equally important aspects of the human mind presents indeed a remarkable analogy to the situation in atomic physics, where complementary phenomena for their definition demand different elementary concepts. Above all, the circumstance that the very word "conscious" refers to experiences capable of being retained in the memory suggests a comparison between conscious experiences and physical observations. In such an analogy, the impossibility of providing an unambiguous content to the idea of subconsciousness corresponds to the impossibility of pictorial interpretation of the quantum-mechanical formalism. Incidentally, psychoanalytical treatment of neuroses may be said to restore balance in the content of the memory of the patient by bringing him new conscious experience, rather than by helping him to fathom the abysses of his subconsciousness.

From a biological point of view, we can only interpret the characteristics of psychical phenomena by concluding that every conscious experience corresponds to a residual impression in the organism, amounting to an irreversible recording in the nervous system of the outcome of processes which are not open to introspection and hardly adapted to exhaustive definition by mechanistic approach. Certainly, such recordings in which the interplay of numerous nerve cells is involved are essentially different from the permanent structures in any single cells of the organism which are connected with genetic reproduction. From a finalistic point of view, however, we may stress not only the usefulness of permanent recordings in their influence on our reactions to subsequent stimuli, but equally the importance that later generations are not encumbered by the actual experiences of individuals but rely only on the reproduction of such properties of the organism as have proved serviceable for the collection and utilization of knowledge. In any attempt to pursue the enquiry we must, of course, be prepared to meet increasing difficulties at every step, and it is suggestive that the simple concepts of physical science lose their immediate applicability to an ever higher degree the more we approach the features of living organisms related to the characteristics of our mind.

To illustrate the argument, we may briefly refer to the old problem of free will. From what has already been said it is evident that the word volition is indispensable to an exhaustive description of psychical phenomena, but the problem is how far we can speak about freedom to act according to our possibilities. As long as unrestricted deterministic views are taken, the idea of such freedom is of course excluded. However, the general lesson of atomic physics, and in particular of the limited scope of mechanistic description of biological phenomena, suggests that the ability of organisms to adjust themselves to environment includes the power of selecting the most appropriate way to this purpose. Because it is impossible to judge such questions on a purely physical basis, it is most important to recognize that psychological experience may offer more pertinent information on the problems. The decisive point is that, if we attempt to predict what another person will decide to do in a given situation, not only must we strive to know his whole background, including the story of his life in all respects which may have contributed to form his character, but we must realize that what we are ultimately aiming at is to put ourselves in his place. Of course, it is impossible to say whether a person wants to do something because he believes he can, or whether he can because he will, but it is hardly disputable that we have the feeling of, so-to-speak, being able to make the best out of the circumstance. From the point of view of objective description, nothing can here be added or taken away, and in this sense we may both practically and logically speak of freedom of will in a way which leaves the proper latitude for the use of words like responsibility and hope, which themselves are as little definable separately as other words indispensable to human communication.

Such considerations point to the epistemological implications of the lesson regarding our observational position, which the development of physical science has impressed upon us. In return for the renunciation of accustomed demands on explanation, it offers a logical means of comprehending wider fields of experience, necessitating proper attention to the placing of the object-subject separation. Since, in philosophical literature, reference is sometimes made to different levels of objectivity or subjectivity or even of reality, it may be stressed that the notion of an ultimate subject as well as conceptions like realism and idealism find no place in objective description as we have defined it; but this circumstance of course does not imply any limitation of the scope of the enquiry with which we are concerned.

Having touched upon some of the problems in science which relate to the unity of knowledge, I shall turn to the further question raised in our programme, whether there is a poetical or spiritual or cultural truth distinct from scientific truth. With all the reluctance of a scientist to enter into such fields, I shall venture, with an attitude similar to that indicated in the preceding, to comment on this question. Taking up the argument of the relation between our means of expression and the field of experience with which we are concerned, we are indeed directly confronted with the relationship of science and art. The enrichment which art can give us originates in its power to remind us of harmonies beyond the grasp of systematic analysis. Literary, pictorial and musical art may be said to form a sequence of modes of expression, where the ever more extensive renunciation of definition, characteristic of scientific communication, leaves fantasy a freer display. In particular, in poetry this purpose is achieved by the juxtaposition of words related to shifting observational situations, thereby emotionally uniting manifold aspects of human knowledge.

Notwithstanding the inspiration required in all work of art, it may not be irreverent to remark that even at the climax of his work the artist relies on the common human foundation on which we stand. In particular, we must realize that a word like improvisation, which comes so readily to the tongue when speaking of artistic achievements, points to a feature essential to all communication. Not only are we in ordinary conversation more or less unaware of the verbal expressions we are going to choose in communicating what is on our minds, but even in written papers, where we have the possibility of reconsidering every word, the question whether to let it stand or change it demands for its answer a final decision essentially equivalent to an improvisation. Incidentally, in the balance between seriousness and humour, characteristic of all truly artistic achievements, we are reminded of complementary aspects conspicuous in children's play and no less appreciated in mature life. Indeed, if we always endeavour to speak quite seriously, we run the risk of very soon appearing ridiculously tedious to our listeners and ourselves, but if we try to joke all the time, we soon find ourselves, and our listeners too, in the desperate mood of the jesters in Shakespeare's dramas.

In a comparison between science and art, we must of course not forget that in the former we have to do with systematic concerted efforts to augment experience and develop appropriate concepts for its comprehension, resembling the carrying and fitting of stones to a building, while in the latter we are presented with more intuitive individual endeavours to evoke sentiments which recall the wholeness of our situation. We are here at a point where the question of unity of knowledge evidently contains ambiguity, like the word "truth" itself. Indeed, with respect to spiritual and cultural values we are also reminded of epistemological problems related to the proper balance between our desire for an all-embracing way of looking at life in its multifarious aspects and our power of expressing ourselves in a logically consistent manner.

Here, essentially different starting points are taken by science, aiming at the development of general methods for ordering common human experience, and religions, originating in endeavours to further harmony of outlook and behaviour within communities. Of course, in any religion, all knowledge shared by the members of the community was included in the general framework, a primary content of which were the values and ideals emphasized in cult and faith. Therefore, the inherent relation between content and frame hardly demanded attention until the subsequent progress of science entailed a novel cosmological· or epistemological lesson. The course of history presents many illustrations in such respects, and we may refer especially to the veritable schism between science and religion which accompanied the development of the mechanical conception of nature at the time of the European Renaissance. On the one hand, many phenomena, hitherto regarded as manifestations of divine providence, appeared as consequences of general immutable laws of nature. On the other hand, the physical methods and viewpoints were far remote from the emphasis on human values and ideals essential to religion. Common to the schools of so-called empirical and critical philosophy, there prevailed therefore an attitude of more or less vague distinction between objective knowledge and subjective belief.

In emphasizing the necessity in unambiguous communication of paying proper attention to the placing of the object-subject separation, modern development of science has, however, created a new basis for the use of such words as knowledge and belief. Above all, the recognition of inherent limitations in the notion of causality has offered a frame in which the idea of universal predestination is replaced by the concept of natural evolution. With respect to the organization of human societies, we may particularly stress that description of the position of the individual within his community presents typically complementary aspects related to the shifting border between the appreciation of values and the background on which they are judged. Surely, every stable human society demands fair play specified in judicial rules, but at the same time, life without attachment to family and friends would obviously be deprived of some of its most precious values. Still, though the closest possible combination of justice and charity presents a common goal in all cultures, it must be recognized that any occasion which calls for the strict application of law has no room for the display of charity and that, conversely, benevolence and compassion may conflict with all ideas of justice. This point, in many religions mythically illustrated by the fight between deities personifying such ideals, is stressed in old Oriental philosophy in the admonition never to forget as we search for harmony in human life that on the scene of existence we are ourselves actors as well as spectators.

In comparing different cultures resting on traditions fostered by historical events, we meet with the difficulty of appreciating the culture of one nation on the background of traditions of another. In this respect, the relation between national cultures has sometimes been described as complementary, although this word cannot here be taken in the strict sense in which it is used in atomic physics or in psychological analysis, where we are dealing with invariable characteristics of our situation. In fact, not only has contact between nations often resulted in the fusion of cultures retaining valuable elements of national traditions, but anthropological research is steadily becoming a most important source for illuminating common features of cultural developments. Indeed, the problem of unity of knowledge can hardly be separated from the striving for universal understanding as a means of elevating human culture.

In concluding this address, I feel that I ought to apologize for speaking on such general topics with so much reference to the special field of knowledge represented by physical science. I have tried, however, to indicate a general attitude suggested by the serious lesson we have in our day received in this field and which to me appears of importance for the problem of unity of knowledge. This attitude may be summarized by the endeavour to achieve a harmonious comprehension of ever wider aspects of our situation, recognizing that no experience is definable without a logical frame and that any apparent disharmony can be removed only by an appropriate widening of the conceptual framework.

Atoms
and
Human Knowledge

1955

In the history of science, this century's exploration of the world of atoms has hardly any parallel in so far as the progress of knowledge and the mastery of that nature of which we ourselves are part are concerned. However, with every increase of knowledge and abilities is connected a greater responsibility; and the fulfilment of the rich promise and the elimination of the new dangers of the atomic age confront our whole civilization with a serious challenge which can be met only by cooperation of all peoples, resting on a mutual understanding of the human fellowship. In this situation, it is important to realize that science, which knows no national boundaries and whose achievements are the common possession of mankind, has through the ages united men in their efforts to elucidate the foundations of our knowledge. As I shall attempt to show, the study of atoms, which was to entail such far-reaching consequences and whose progress has been based on world-wide cooperation, not only has deepened our insight into a new domain of experience, but has thrown new light on general problems of knowledge.

At first, it might seem surprising that atomic science should contain a lesson of a general nature, but we must remember that it has in all stages of its development concerned profound problems of knowledge. Thus, thinkers of antiquity, by assuming a limit for the divisibility of substances, attempted to find a basis for understanding the features of permanency exhibited by natural phenomena, in spite of their multifariousness and variability. Although atomic ideas have contributed more and more fruitfully to the development of physics and chemistry since the Renaissance, they were regarded as a hypothesis right up to the beginning of this century. Indeed, it was taken for granted that our sense organs, themselves composed of innumerable atoms, were too coarse to observe the smallest parts of matter. This situation was, however, to become essentially changed by the great discoveries at the turn of the century and, as is well known, progress in experimental technique made it possible to record the effects of single atoms and to obtain information on the more elementary particles of which the atoms themselves were found to be composed.

In spite of the deep influence exerted by ancient atomism on the development of the mechanical conception of nature, it was the study of immediately accessible astronomical and physical experience which made it possible to trace the regularities expressed in the so-called classical physics. Galileo's dictum, according to which the account of phenomena should be based on measurable quantities, made it possible to eliminate such animistic views which had so long hindered the rational formulation of mechanics. In Newton's principles, the foundation was laid of a deterministic description permitting, from the knowledge of the state of a physical system at a given moment, prediction of its state at any subsequent time. On the same lines, it was possible to account for electromagnetic phenomena. This required, however, that the description of the state of the system should include, besides the positions and velocities of the electrified and magnetized bodies, the strength and direction of the electrical and magnetic forces at every point of space, at the given moment.

The conceptual framework which is characteristic of classical physics was long thought to provide the correct tool for the description of all physical phenomena, and not least was it suited to the utilization and development of atomic ideas. Of course, for systems, such as ordinary bodies which are composed of an enormous number of constituent parts, there could be no question of an exhaustive description of the state of the system. Without abandoning the deterministic ideal, it became possible, however, on the basis of the principles of classical mechanics, to deduce statistical regularities reflecting many of the properties of material bodies. Even though the mechanical laws of motion permit a complete reversal of the course of single processes, full explanation of the characteristic feature of irreversibility in heat phenomena was found in the statistical energy equilibrium resulting from the interaction of the molecules. This great extension of the application of mechanics emphasized further the indispensability of atomic ideas to the description of nature and opened the first possibilities of counting the atoms of the substances.

However, clarification of the foundation of the laws of thermodynamics was to open the way for recognition of a feature of wholeness in atomic processes far beyond the old doctrine of the limited divisibility of matter. As is well known, the closer analysis of heat radiation became the test of the scope of classical physical ideas. The discovery of electromagnetic waves had already provided a basis for understanding the propagation of light, explaining many of the optical properties of substances; but endeavours to account for radiation equilibrium confronted such ideas with insurmountable difficulties. The circumstance that one had here to do with arguments based on general principles and quite independent of special assumptions regarding the constituents of the substances led Planck, in the first year of this century, to the discovery of the universal quantum of action, which showed clearly that the classical physical description is an idealization of limited applicability. In phenomena on the ordinary scale, the actions involved are so large compared to the quantum that it can be left out of consideration. However, in proper quantum processes, we meet regularities which are completely foreign to the mechanical conception of nature and which defy pictorial deterministic description.

The task with which Planck's discovery confronted physicists was nothing less than, by means of a thorough analysis of the presuppositions on which the application of our most elementary concepts are based, to provide room for the quantum of action in a rational generalization of the classical physical description. During the development of quantum physics, entailing so many surprises, we have time and again been reminded of the difficulties of orienting ourselves in a domain of experience far from that to the description of which our means of expression are adapted. Rapid progress has been made possible by a wide and intensive collaboration among physicists from many countries, whose diverse approaches have helped in a most fruitful way to focus the problem ever more sharply. On this occasion, of course, it will not be possible to deal in detail with individual contributions, but as a background for the following considerations I shall remind you briefly of some of the main features of the development.

While Planck cautiously limited himself to statistical arguments and emphasized the difficulties of abandoning the classical foundations in the detailed description of nature, Einstein daringly pointed to the necessity of taking the quantum of action into account in individual atomic phenomena. In the same year that he so harmoniously completed the framework of classical physics by establishing the theory of relativity, he showed that the description of observations on photoelectric effects requires that the transmission of energy to each of the electrons expelled from the substances corresponds to the absorption of a so-called quantum of radiation. Since the idea of waves is indispensable to the account of the propagation of light, there could be no question of simply replacing it with a corpuscular description, and one was therefore confronted with a peculiar dilemma whose solution was to require a thorough analysis of the scope of pictorial concepts.

As is well known, this question was further accentuated by Rutherford's discovery of the atomic nucleus which, despite its minuteness, contains almost the whole mass of the atom and whose electrical charge corresponds to the number of electrons in the neutral atom. This gave a simple picture of the atom which immediately suggested the application of mechanical and electromagnetic ideas. Yet, it was clear that, according to classical physical principles, no configuration of electrical particles could possess the stability necessary to the explanation of the physical and chemical properties of atoms. In particular, according to classical electromagnetic theory, every motion of the electrons around the atomic nucleus would produce a continual radiation of energy implying a rapid contraction of the system until the electrons became united with the nucleus into a neutral particle of dimensions vanishingly small relative to those which must be ascribed to atoms. However, in the hitherto entirely incomprehensible empirical laws for the line spectra of the elements was found a hint as to the decisive importance of the quantum of action for the stability and radiative reactions of the atom.

The point of departure became here the so-called quantum postulate, according to which every change in the energy of an atom is the result of a complete transition between two of its stationary states. By assuming further that all atomic radiative reactions involve the emission or absorption of a single light quantum, the energy values of the stationary states could be determined from the spectra. It was evident that no explanation of the indivisibility of the transition processes, or their appearance under given conditions, could be given within the framework of deterministic description. However, it proved possible to obtain a survey of the electron bindings in the atom, which reflected many of the properties of substances, with the aid of the so-called correspondence principle. On the basis of a comparison with the classically expected course of the processes, directives were sought for a statistical generalization of the description compatible with the quantum postulate. Still, it became more and more clear that, in order to obtain a consistent account of atomic phenomena, it was necessary to renounce even more the use of pictures and that a radical reformulation of the whole description was needed to provide room for all features implied by the quantum of action.

The solution which was reached as a result of the ingenious contributions of many of the most eminent theoretical physicists of our time was surprisingly simple. As in the formulation of relativity theory, adequate tools were found in highly developed mathematical abstractions. The quantities which in classical physics are used to describe the state of a system are replaced in quantum-mechanical formalism by symbolic operators whose commutability is limited by rules containing the quantum. This implies that quantities such as positional coordinates and corresponding momentum components of particles cannot simultaneously be ascribed definite values. In this way, the statistical character of the formalism is displayed as a natural generalization of the description of classical physics. In addition, this generalization permitted a consequent formulation of the regularities which limit the individuality of identical particles and which, like the quantum itself, cannot be expressed in terms of usual physical pictures.

By means of the methods of quantum mechanics it was possible to account for a very large amount of the experimental evidence on the physical and chemical properties of substances. Not only was the binding of electrons in atoms and molecules clarified in detail, but a deep insight was also obtained into the constitution and reactions of atomic nuclei. In this connection, we may mention that the probability laws for spontaneous radioactive transmutations have been harmoniously incorporated into the statistical quantum-mechanical description. Also the understanding of the properties of the new elementary particles, which have been observed in recent years in the study of transmutations of atomic nuclei at high energies, has been subject to continual progress resulting from the adaption of the formalism to the invariance requirements of relativity theory. Still, we are here confronted with new problems whose solution obviously demands further abstractions suited to combine the quantum of action with the elementary electric charge.

In spite of the fruitfulness of quantum mechanics within such a wide domain of experience, the renunciation of accustomed demands on physical explanation has caused many physicists and philosophers to doubt that we are here dealing with an exhaustive description of atomic phenomena. In particular, the view has been expressed that the statistical mode of description must be regarded as a temporary expedient which, in principle, ought to be replaceable by a deterministic description. The thorough discussion of this question has, however, led to that clarification of our position as observers in atomic physics which has given us the epistemological lesson referred to in the beginning of this lecture.

As the goal of science is to augment and order our experience, every analysis of the conditions of human knowledge must rest on considerations of the character and scope of our means of communication. Our basis is, of course, the language developed for orientation in our surroundings and for the organization of human communities. However, the increase of experience has repeatedly raised questions as to the sufficiency of the concepts and ideas incorporated in daily language. Because of the relative simplicity of physical problems, they are especially suited to investigate the use of our means of communication. Indeed, the development of atomic physics has taught us how, without leaving common language, it is possible to create a framework sufficiently wide for an exhaustive description of new experience.

In this connection, it is imperative to realize that in every account of physical experience one must describe both experimental conditions and observations by the same means of communication as one used in classical physics. In the analysis of single atomic particles, this is made possible by irreversible amplification effects—such as a spot on a photographic plate left by the impact of an electron, or an electric discharge created in a counter device—and the observations concern only where and when the particle is registered on the plate or its energy on arrival at the counter. Of course, this information presupposes knowledge of the position of the photographic plate relative to the other parts of the experimental arrangement, such as regulating diaphragms and shutters defining space-time coordination or electrified and magnetized bodies which determine the external force fields acting on the particle and permit energy measurements. The experimental conditions can be varied in many ways, but the point is that in each case we must be able to communicate to others what we have done and what we have learned, and that therefore the functioning of the measuring instruments must be described within the framework of classical physical ideas.

As all measurements thus concern bodies sufficiently heavy to permit the quantum to be neglected in their description, there is, strictly speaking, no new observational problem in atomic physics. The amplification of atomic effects, which makes it possible to base the account on measurable quantities and which gives the phenomena a peculiar closed character, only emphasizes the irreversibility characteristic of the very concept of observation. While, within the frame of classical physics, there is no difference in principle between the description of the measuring instruments and the objects under investigation, the situation is essentially different when we study quantum phenomena, since the quantum of action imposes restrictions on the description of the state of the systems by means of space-time coordinates and momentum-energy quantities. Since the deterministic description of classical physics rests on the assumption of an unrestricted compatibility of space-time coordination and the dynamical conservation laws, we are obviously confronted here with the problem of whether, as regards atomic objects, such a description can be fully retained.

The role of the interaction between objects and measuring instruments in the description of quantum phenomena was found to be especially important for the clarification of this main point. Thus, as stressed by Heisenberg, the locating of an object in a limited space-time domain involves, according to quantum mechanics, an exchange of momentum and energy between instrument and object which is the greater the smaller the domain chosen. It was therefore of the utmost importance to investigate the extent to which the interaction entailed in observation can be taken into account separately in the description of phenomena. This question has been the focus of much discussion, and there have appeared many proposals which aim at the complete control of all interactions. In such considerations, however, due regard is not taken to the fact that the very account of the functioning of measuring instruments involves that any interaction, implied by the quantum, between these and the atomic objects, be inseparably entailed in the phenomena.

Indeed, every experimental arrangement permitting the registration of an atomic particle in a limited space-time domain demands fixed measuring rods and synchronized clocks which, from their very definition, exclude the control of momentum and energy transmitted to them. Conversely, any unambiguous application of the dynamical conservation laws in quantum physics requires that the description of the phenomena involve a renunciation in principle of detailed space-time coordination. This mutual exclusiveness of the experimental conditions implies that the whole experimental arrangement must be taken into account in a well-defined description of the phenomena. The indivisibility of quantum phenomena finds its consequent expression in the circumstance that every definable subdivision would require a change of the experimental arrangement with the appearance of new individual phenomena. Thus, the very foundation of a deterministic description has disappeared and the statistical character of the predictions is evidenced by the fact that in one and the same experimental arrangement there will in general appear observations corresponding to different individual processes.

Such considerations not only have clarified the above-mentioned dilemma with respect to the propagation of light, but have also completely solved the corresponding paradoxes confronting pictorial representation of the behaviour of material particles. Here, of course, we cannot seek a physical explanation in the customary sense, but all we can demand in a new field of experience is the removal of any apparent contradiction. However great the contrasts exhibited by atomic phenomena under different experimental conditions, such phenomena must be termed complementary in the sense that each is well defined and that together they exhaust all definable knowledge about the objects concerned. The quantum-mechanical formalism, the sole aim of which is the comprehension of observations obtained under experimental conditions described by simple physical concepts, gives just such an exhaustive complementary account of a very large domain of experience. The renunciation of pictorial representation involves only the state of atomic objects, while the foundation of the description of the experimental conditions, as well as our freedom to choose them, is fully retained. The whole formalism which can be applied only to closed phenomena must in all such respects be considered a rational generalization of classical physics.

In view of the influence of the mechanical conception of nature on philosophical thinking, it is understandable that one has sometimes seen in the notion of complementarity a reference to the subjective observer, incompatible with the objectivity of scientific description. Of course, in every field of experience we must retain a sharp distinction between the observer and the content of the observations, but we must realize that the discovery of the quantum of action has thrown new light on the very foundation of the description of nature and revealed hitherto unnoticed presuppositions to the rational use of the concepts on which the communication of experience rests. In quantum physics, as we have seen, an account of the functioning of the measuring instruments is indispensable to the definition of phenomena and we must, so-to-speak, distinguish between subject and object in such a way that each single case secures the unambiguous application of the elementary physical concepts used in the description. Far from containing any mysticism foreign to the spirit of science, the notion of complementarity points to the logical conditions for description and comprehension of experience in atomic physics.

The epistemological lesson of atomic physics has naturally, just as have earlier advances in physical science, given rise to renewed consideration of the use of our means of communication for objective description in other fields of knowledge. Not least the emphasis placed on the observational problem raises the question of the position of living organisms in the description of nature and of our own situation as thinking and acting beings. Even though it was, to some extent, possible within the frame of classical physics to compare organisms with machines, it was clear that such comparisons did not take sufficient account of many of the characteristics of life. The inadequacy of the mechanical concept of nature for the description of man's situation is particularly evident in the difficulties entailed in the primitive distinction between soul and body.

The problems with which we are confronted here are obviously connected with the fact that the description of many aspects of human existence demands a terminology which is not immediately founded on simple physical pictures. However, recognition of the limited applicability of such pictures in the account of atomic phenomena gives a hint as to how biological and psychological phenomena may be comprehended within the frame of objective description. As before, it is here important to be aware of the separation between the observer and the content of the communications. While, in the mechanical conception of nature, the subject-object distinction was fixed, room is provided for a wider description through the recognition that the consequent use of our concepts requires different placings of such a separation.

Without attempting any exhaustive definition of organic life, we may say that a living organism is characterized by its integrity and adaptability, which implies that a description of the internal functions of an organism and its reaction to external stimuli often requires the word purposeful, which is foreign to physics and chemistry. Although the results of atomic physics have found a multitude of applications in biophysics and biochemistry, the closed individual quantum phenomena exhibit, of course, no feature suggesting the notion of life. As we have seen, the description of atomic phenomena, exhaustive within a wide domain of experience, is based on the free use of such measuring instruments as are necessary to proper application of the elementary concepts. In a living organism, however, such a distinction between the measuring instruments and the objects under investigation can hardly be fully carried through, and we must be prepared that every experimental arrangement whose aim is a description of the functioning of the organism, which is well defined in the sense of atomic physics, will be incompatible with the display of life.

In biological research, references to features of wholeness and purposeful reactions of organisms are used together with the increasingly detailed information on structure and regulatory processes that has resulted in such great progress not least in medicine. We have here to do with a practical approach to a field where the means of expression used for the description of its various aspects refer to mutually exclusive conditions of observation. In this connection, it must be realized that the attitudes termed mechanistic and finalistic are not contradictory points of view, but rather exhibit a complementary relationship which is connected with our position as observers of nature. To avoid misunderstanding, however, it is essential to note that—in contrast to the account of atomic regularities—a description of organic life and an evaluation of its possibilities of development cannot aim at completeness, but only at sufficient width of the conceptual framework.

In the account of psychical experiences, we meet conditions of observation and corresponding means of expression still further removed from the terminology of physics. Quite apart from the extent to which the use of words like instinct and reason in the description of animal behaviour is necessary and justifiable, the word consciousness, applied to oneself as well as to others, is indispensable when describing the human situation. While the terminology adapted to orientation in the environment could take as its starting point simple physical pictures and ideas of causality, the account of our states of mind required a typical complementary mode of description.

Indeed, the use of words like thought and feeling does not refer to a firmly connected causal chain, but to experiences which exclude each other because of different distinctions between the conscious content and the background which we loosely term ourselves. The relation between the experience of a feeling of volition and conscious pondering on motives for action is especially instructive. The indispensability of such apparently contrasting means of expression to the description of the richness of conscious life strikingly reminds us of the way in which elementary physical concepts are used in atomic physics. In such a comparison, however, we must recognize that psychical experience cannot be subjected to physical measurements and that the very concept of volition does not refer to a generalization of a deterministic description, but from the outset points to characteristics of human life. Without entering into the old philosophical discussion of freedom of the will, I shall only mention that in an objective description of our situation use of the word volition corresponds closely to that of words like hope and responsibility, which are equally indispensable to human communication.

We have here reached problems which touch human fellowship and where the variety of means of expression originates from the impossibility of characterizing by any fixed distinction the role of the individual in the society. The fact that human cultures, developed under different conditions of living, exhibit such contrasts with respect to established traditions and social patterns allows one, in a certain sense, to call such cultures complementary. However, we are here in no way dealing with definite mutually exclusive features, such as those we meet in the objective description of general problems of physics and psychology, but with differences in attitude which can be appreciated or ameliorated by extended intercourse between peoples. In our time, when increasing knowledge and ability more than ever link the fate of all peoples, international collaboration in science has far-reaching tasks which may be furthered not least by an awareness of the general conditions for human knowledge.

Physical Science
and
the Problem of Life

1957

It has been a pleasure to accept the invitation of the Medical Society of Copenhagen to give one of the Steno lectures by which the Society commemorates the famous Danish scientist whose achievements are admired in ever greater measure, not only in this country but in the whole scientific world. As my theme I have chosen a problem which has occupied the human mind through the ages and with which Niels Stensen himself was deeply concerned, namely how far physical experience can help us to explain organic life in its rich and varied display. As I shall try to show, the development of physics in recent decades and in particular the lesson regarding our position as observers of that nature of which we are part, received through the exploration of the world of atoms so long closed to us, have created a new background for our attitude to this question.

Even in the philosophical schools of ancient Greece, we find divergent opinions regarding the conceptual means suited to account for the striking differences between living organisms and other material bodies. As is well known, the atomists considered a limited divisibility of all matter necessary to explain not only simple physical phenomena but also the functioning of organisms and the related psychical experiences. Aristotle, on the other hand, refuted atomic ideas and, in view of the wholeness exhibited by every living organism, maintained the necessity of introducing into the description of nature such concepts as perfection and purposefulness.

For almost 2000 years the situation remained essentially unchanged, and not until the Renaissance did there occur the great discoveries in physics as well as in biology, which were to give new incentives. Progress in physics consisted above all in the liberation from the Aristotelian idea of driving forces as the cause of all motion. Galileo's recognition that uniform motion is a manifestation of inertia and his emphasis on force as a cause of change of motion were to become the foundation of the development of mechanics, which Newton to the admiration of succeeding generations endowed with a firm and completed form. In this so-called classical mechanics all reference to purpose is eliminated, since the course of events is described as automatic consequences of given initial conditions.

The progress of mechanics could not avoid making the strongest impression on all contemporary science. In particular, the anatomical studies of Vesalius and Harvey's discovery of the circulation of the blood suggested the comparison between living organisms and machines working according to the laws of mechanics. On the philosophical side, it was especially Descartes who stressed the similarity between animals and automata, but ascribed to human beings a soul interacting with the body in a certain gland in the brain. However, the insufficiency of contemporary knowledge of such problems was emphasized by Steno in his famous Paris lecture on the anatomy of the brain, which bears witness of the great observational power and open-mindedness characteristic of all Steno's scientific work.

Subsequent developments in biology, especially after the invention of the microscope, revealed an unsuspected fineness in organic structure and regulatory processes. At the same time that mechanistic ideas thus found ever wider applications, so-called vitalistic or finalistic points of view, inspired by the wonderful power of regeneration and adaptation in organisms, were repeatedly expressed. Rather than returning to primitive ideas of a life force acting in the organisms, such views emphasized the insufficiency of physical approach in accounting for the characteristics of life. As a sober presentation of the situation as it stood in the beginning of this century, I should like to refer to the following statement by my father, the physiologist Christian Bohr, in the introduction to his paper "On Pathological Lung Expansion" which appeared in the anniversary publication of the Copenhagen University in 1910.

As far as physiology can be characterized as a special branch of natural sciences, its specific task is to investigate the phenomena peculiar to the organism as a given empirical object in order to obtain an understanding of the various parts in the self-regulation and how they are balanced against each other and brought into harmony with variations in external influences and inner processes. It is thus in the very nature of this task to refer the word purpose to the maintenance of the organism and consider as purposive the regulation mechanisms which serve this maintenance. Just in this sense we shall in the following use the notion "purposiveness" about organic functions. In order that the application of this concept in each single case should not be empty or even misleading it must, however, be demanded that it be always preceded by an investigation of the organic phenomenon under consideration, sufficiently thorough to illuminate step by step the special way in which it contributes to the maintenance of the organism. Although this demand, which requires no more than the scientific demonstration that the notion of purposiveness in the given case is used in accordance with its definition, might appear self-evident, it may nevertheless not be unnecessary to stress it. Indeed, physiological investigations have brought to light regulations of utmost fineness in a multitude so great that it is a temptation to designate every observed manifestation of life as purposive without attempting an experimental investigation of its detailed functioning. By means of analogies which so easily present themselves among the variety of organic functions, it is merely the next step to interpret this functioning from a subjective judgement about the special character of purposiveness in the given case. It is evident, however, how often, with our so narrowly limited knowledge about the organism, such a personal judgement may be erroneous, as is illustrated by many examples. In such cases, it is the lacking experimental illumination of the details of the process which is the cause of the erroneous results of the procedure. The *a priori* assumption of the purposiveness of the organic process is, however, in itself quite natural as a heuristic principle and can, due to the extreme complication and difficult comprehension of the conditions in the organism, prove not only useful, but even indispensable for the formulation of the special problem for the investigation and the search of ways for its solution. But one thing is what may be conveniently used by the preliminary investigation, another what justifiably can be considered an actually achieved result. As regards the problem of the purposiveness of a given function for the maintenance of the whole organism, such a result can, as stressed above, be secured only by a demonstration in detail of the ways in which the purpose is reached.

I have quoted these remarks which express the attitude in the circle in which I grew up and to whose discussions I listened in my youth, because they offer a suitable starting point for the investigation of the place of living organisms in the description of nature. As I shall try to show, modern development of atomic physics, at the same time as it has augmented our knowledge about atoms and their constitution of more elementary parts, has revealed the limitation in principle of the so-called mechanical conception of nature and thereby created a new background for the problem, most pertinent to our subject, as to what we can understand by and demand of a scientific explanation.

In order to present the situation in physics as clearly as possible, I shall start by reminding you of the extreme attitude which, under the impact of the great success of classical mechanics, was expressed in Laplace's well-known conception of a world machine. All interactions between the constituents of this machine were governed by the laws of mechanics, and therefore an intelligence knowing the relative positions and velocities of these parts at a given moment could predict all the subsequent events in the world, including the behaviour of animals and man. In this whole conception which has, as is well known, played an important role in philosophical discussion, due attention is not paid to the presuppositions for the applicability of the concepts indispensable for communication of experience.

In this respect the later development of physics has given us an urgent lesson. Already the far-reaching interpretation of heat phenomena as incessant motion of molecules in gases, liquids, and solids has called attention to the importance of the conditions of observation in the account of experience. Of course, there could be no question of a detailed description of the motions of the innumerable particles among each other, but only of deducing statistical regularities of heat motion by means of general mechanical principles. The peculiar contrast between the reversibility of simple mechanical processes and the irreversibility typical of many thermodynamical phenomena was thus clarified by the fact that applications of such concepts as temperature and entropy refer to experimental conditions incompatible with complete control of the motions of single molecules.

In the maintenance and growth of living organisms one has often seen a contradiction to that tendency, implied by the thermodynamical laws, towards temperature and energy equilibrium in an isolated physical system. However, we must remember that organisms are continually supplied with free energy by nutrition and respiration, and the most thorough physiological investigation has never revealed any departure from thermodynamical principles. Yet, recognition of such similarities between living organisms and ordinary power engines is of course in no way sufficient to answer the question about the position of organisms in the description of nature, a question obviously demanding deeper analysis of the observational problem.

This very problem has indeed been brought to the foreground in an unexpected way by the discovery of the universal quantum of action which expresses a feature of wholeness in atomic processes that prevents the distinction between observation of phenomena and independent behaviour of the objects, characteristic of the mechanical conception of nature. In physical systems on the ordinary scale, representation of events as a chain of states described by measurable quantities rests on the circumstance that all actions are here large enough to permit neglect of the interaction between the objects and the bodies used as measuring tools. Under conditions where the quantum of action plays a decisive part and where such an interaction is therefore an integral part of the phenomena, there cannot to the same extent be ascribed a mechanically well-defined course.

The breakdown of ordinary physical pictures which here confronts us is strikingly expressed in the difficulties in talking about properties of atomic objects independent of the conditions of observation. Indeed, an electron may be called a charged material particle, since measurements of its inertial mass always give the same result, and any transmission of electricity between atomic systems always amounts to a number of so-called unit charges. Yet, the interference effects appearing when electrons pass through crystals are incompatible with the mechanical idea of particle motion. We meet analogous features in the well-known dilemma about the nature of light, since optical phenomena require the notion of wave propagation, while the laws of transmission of momentum and energy in atomic photo-effects refer to the mechanical conception of particles.

This situation, novel in physical science, has demanded a renewed analysis of the presuppositions for the application of concepts used for orientation in our surroundings. Of course, in atomic physics we retain the freedom by experimenting to put questions to nature, but we must recognize that the experimental conditions which can be varied in numerous ways are defined only by bodies so heavy that in the description of their functions we can disregard the quantum. Information concerning atomic objects consists solely in the marks they make on these measuring instruments, as, for instance, a spot produced by the impact of an electron on a photographic plate placed in the experimental arrangement. The circumstance that such marks are due to irreversible amplification effects endows the phenomena with a peculiarly closed character pointing directly to the irreversibility in principle of the very notion of observation.

The special situation in quantum physics is above all, however, that the information gained about atomic objects cannot be comprehended along the lines of approach typical of the mechanical conception of nature. Already the fact that under one and the same experimental arrangement there may in general appear observations pertaining to different individual quantum processes entails a limitation in principle of the deterministic mode of description. The demand of unrestricted divisibility on which classical physical description rests is also clearly incompatible with that feature of wholeness in typical quantum phenomena which involves that any definable subdivision requires a change of the experimental arrangement giving rise to new individual effects.

In order to characterize the relation between phenomena observed under different experimental conditions, one has introduced the term complementarity to emphasize that such phenomena together exhaust all definable information about the atomic objects. Far from containing any arbitrary renunciation of customary physical explanation, the notion of complementarity refers directly to our position as observers in a domain of experience where unambiguous application of the concepts used in the description of phenomena depends essentially on the conditions of observation. By a mathematical generalization of the conceptual framework of classical physics it has been possible to develop a formalism which leaves room for the logical incorporation of the quantum of action. This so-called quantum mechanics aims directly at the formulation of statistical regularities pertaining to evidence gained under well-defined observational conditions. The completeness in principle of this description is due to the retention of classical mechanical ideas to an extent including any definable variation of the experimental conditions.

The complementary character of the quantum-mechanical description is clearly expressed in the account of the composition and reactions of atomic systems. Thus, the regularities regarding the energy states of atoms and molecules, responsible for the characteristic spectra of the elements and the valences of chemical combinations, appear only under circumstances where a control of the positions of the electrons within the atom and the molecule is excluded. In this connection, it is interesting to note that fruitful application of the structural formulae in chemistry rests solely on the fact that the atomic nuclei are so much heavier than the electrons. However, with respect to the stability and transmutations of the atomic nuclei themselves, quantum-mechanical features are again decisive. Only in a complementary description transcending the scope of the mechanical conception of nature is it possible to find room for the fundamental regularities responsible for the properties of the substances of which our tools and our bodies are composed.

Progress in the field of atomic physics has, as is well known, found wide application in the biological sciences. In particular, I may mention the understanding we have gained of the peculiar stability of chemical structures in the cells responsible for the hereditary properties of the species, and of the statistical laws for the occurrence of mutations produced by exposing organisms to special agencies. Furthermore, amplification effects similar to those permitting observation of individual atomic particles play a decisive role in many functions of the organism. In this way is stressed the irreversible character of typical biological phenomena, and the time direction inherent in the description of the functioning of organisms is strikingly marked by their utilization of past experience for reactions to future stimuli.

In this promising development we have to do with a very important and, according to its character, hardly limited extension of the application of purely physical and chemical ideas to biological problems, and since quantum mechanics appears as a rational generalization of classical physics, the whole approach may be termed mechanistic. The question, however, is in what sense such progress has removed the foundation for the application of so-called finalistic arguments in biology. Here we must realize that the description and comprehension of the closed quantum phenomena exhibit no feature indicating that an organization of atoms is able to adapt itself to the surroundings in the way we witness in the maintenance and evolution of living organisms. Furthermore, it must be stressed that an account, exhaustive in the sense of quantum physics, of all the continually exchanged atoms in the organism not only is infeasible but would obviously require observational conditions incompatible with the display of life.

However, the lesson with respect to the role which the tools of observation play in defining the elementary physical concepts gives a clue to logical application of notions like purposiveness foreign to physics, but lending themselves so readily to the description of organic phenomena. Indeed, on this background it is evident that the attitudes termed mechanistic and finalistic do not present contradictory views on biological problems, but rather stress the mutually exclusive character of observational conditions equally indispensable in our search for an ever richer description of life. Here, there is of course no question of an explanation akin to the classical physical account of the functioning of simple mechanical constructions or of complicated electron calculation machines, but we are concerned with wider pursuit of that analysis of the presuppositions and scope of our conceptual means of communication which has become so characteristic of the newer development of physics.

Apart from all differences with respect to observational conditions, communication of biological experiences contains no more reference to the subjective observer than does the description of physical evidence. Thus, so far it has not been necessary to enter more closely into the conditions of observation characteristic for the account of psychological phenomena, for which we cannot rely on the conceptual frame developed for our orientation in inanimate nature. However, the fact that conscious experience can be remembered and therefore must be supposed to be connected with permanent changes in the constitution of the organism points to a comparison between psychical experiences and physical observations. With respect to relationships between conscious experiences we also encounter features reminiscent of the conditions for the comprehension of atomic phenomena. The rich vocabulary used in the communications of the states of our mind refers indeed to a typical complementary mode of description corresponding to a continual change of the content on which attention is focused.

Compared to the extension of the mechanical mode of description demanded by the account of the individuality of atomic phenomena, the integrity of the organism and the unity of the personality confront us of course with a further generalization of the frame for the rational use of our means of communication. In this respect, it must be emphasized that the distinction between subject and object, necessary for unambiguous description, is retained in the way that in every communication containing a reference to ourselves we, so-to-speak, introduce a new subject which does not appear as part of the content of the communication. It need hardly be stressed that it is just this freedom of choosing the subject-object distinction which provides room for the multifariousness of conscious phenomena and the richness of human life.

The attitude to general problems of knowledge to which the development of physics in this century has led us differs essentially from the approach to such problems at Steno's time. This does not mean, however, that we have left the road to the enrichment of knowledge followed by him with such great results, but we have realized that the striving for beauty and harmony which marked Steno's work demands a steady revision of the presuppositions and scope of our means of communication.

complicated electron calculation machines, but we are concerned with wider pursuit of that analysis of the presuppositions and scope of our conceptual means of communication which has become so characteristic of the newer development of physics.

Apart from all differences with respect to observational conditions, communication of biological experiences contains no more reference to the subjective observer than does the description of physical evidence. Thus, so far it has not been necessary to enter more closely into the conditions of observation characteristic for the account of psychological phenomena, for which we cannot rely on the conceptual frame developed for our orientation in inanimate nature. However, the fact that conscious experience can be remembered and therefore must be supposed to be connected with permanent changes in the constitution of the organism points to a comparison between psychical experiences and physical observations. With respect to relationships between conscious experiences we also encounter features reminiscent of the conditions for the comprehension of atomic phenomena. The rich vocabulary used in the communications of the states of our mind refers indeed to a typical complementary mode of description corresponding to a continual change of the content on which attention is focused.

Compared to the extension of the mechanical mode of description demanded by the account of the individuality of atomic phenomena, the integrity of the organism and the unity of the personality confront us of course with a further generalization of the frame for the rational use of our means of communication. In this respect, it must be emphasized that the distinction between subject and object, necessary for unambiguous description, is retained in the way that in every communication containing a reference to ourselves we, so-to-speak, introduce a new subject which does not appear as part of the content of the communication. It need hardly be stressed that it is just this freedom of choosing the subject-object distinction which provides room for the multifariousness of conscious phenomena and the richness of human life.

The attitude to general problems of knowledge to which the development of physics in this century has led us differs essentially from the approach to such problems at Steno's time. This does not mean, however, that we have left the road to the enrichment of knowledge followed by him with such great results, but we have realized that the striving for beauty and harmony which marked Steno's work demands a steady revision of the presuppositions and scope of our means of communication.